RAMAN SPECTROSCOPY
An Intensity Approach

RAMAN SPECTROSCOPY

An Intensity Approach

Wu Guozhen
Tsinghua University, China

Published by

World Scientific Publishing Co. Pte. Ltd.

5 Toh Tuck Link, Singapore 596224

USA office: 27 Warren Street, Suite 401-402, Hackensack, NJ 07601

UK office: 57 Shelton Street, Covent Garden, London WC2H 9HE

Library of Congress Cataloging-in-Publication Data
Names: Wu, Guozhen, 1947– author.
Title: Raman spectroscopy : an intensity approach / Wu Guozhen, Tsinghua University, China.
Description: Hackensack, NJ : World Scientific, [2016] |
Includes bibliographical references and index.
Identifiers: LCCN 2016025842| ISBN 9789813143494 (hardcover ; alk. paper) |
ISBN 9813143495 (hardcover ; alk. paper)
Subjects: LCSH: Raman spectroscopy. | Raman effect.
Classification: LCC QC454.R36 W8 2016 | DDC 535.8/46--dc23
LC record available at https://lccn.loc.gov/2016025842

British Library Cataloguing-in-Publication Data
A catalogue record for this book is available from the British Library.

Desk Editor: Christopher Teo

Typeset by Stallion Press
Email: enquiries@stallionpress.com

Printed in Singapore

Preface

I began to recognize the importance of studying Raman intensity as I encountered surface enhanced Raman scattering (SERS) back to 1980's. SERS is a surface phenomenon that as a molecule, especially the nitrogen containing molecule, is adsorbed on the metal surface, in particular the silver electrode, its Raman cross section can be amplified up to a million fold. More interesting is that the Raman mode intensities of the adsorbed molecule are applied voltage dependent. The question then in my mind was: what is the physical picture behind this Raman intensity variation? In order to solve this issue, we therefore established an algorithm to retrieve the so-called bond polarizabilities from the Raman mode intensities in a systematic way. This leads to a nice harvest, showing that this approach is an adequate direction, albeit the algorithm is semi-classical. This attracted much of my intention in 1980's.

During that time, I also thought of the fields of Raman optical activity (ROA) and phase transition. These two fields involve Raman intensity variation as well. In particular, ROA shows that for a chiral molecule under right and left circularly polarized light scatterings, its respective Raman intensities are different, though the difference is very small, only 10^{-3} to 10^{-4} of its Raman intensity. The differential Raman intensity is called the ROA spectrum.

Our work on the phase transitions of the systems with very low degree of doping (ranging from 10^{-2} to 10^{-4}) seemed to be a success. The rate of mode intensity variation as a function of temperature shows a power law. The exponent of the power law is very sensitive to the doping degree and bears the information of the doping effects. Among the doping effects, the self-similarity by doping which is characterized by the scaling factor, $d/\sqrt{M}$, with d the separation between the doping ions and M their mass is most impressive.

However, our work on ROA turned out to be a maze, but not a loss. I hence was acquainted with the idea of ROA and proposed a classical formula for predicting the ROA mode signatures①. Though with this formula, the prediction was not so successful due to the reason that, at that time, we did not have a clear picture concerning the Raman excited virtual state from the retrieved bond polarizabilities.

A clear picture of the Raman excited virtual state sparked us in 2006 when we noticed that the bond polarizabilities retrieved from Raman mode intensities were definitely in variation with the bond electronic densities in the ground state②. This hints that bond polarizabilities bear the information of the excited/disturbed charges during the Raman process, i. e. , the electronic structure of the Raman virtual state! With this breakthrough, we immediately came into the detailed study on the Raman virtual state and extended the bond polarizability algorithm to retrieve the differential bond polarizabilities from ROA mode intensities. All these offered us vivid pictures of the Raman and ROA processes. Though they are classical pictures, just because they are classical, they can provide comprehensive pictures of the Raman virtual state and the phenomena in ROA, which were not known or neglected before. Furthermore, our classical formula for ROA which was developed in 1998, turns out to be a success after its re-interpretation by the bond polarizability. Another byproduct of this work is that we showed, probably for the first time to the best of our knowledge, that there are about 20% electrons in a molecule that are involved in the Raman process③. The 14 years' waiting (from 1998 to 2012) is really worthwhile to me!

My bias is that, at least in the field of molecular physics or molecular spectroscopy, quantum treatment is often a requisite, but not an absolute necessity. Classical or semi-classical treatment is in ge-

① Ma S, Wu G. Chinese Phys. Letters, 1998, 15: 753

② See Fig. 4. 3 of Chapter 4 and Wang H, Wu G. Chem. Phys. Letters, 2006, 421: 460

③ Fang Y, Wu G, Wang P. Spectrochimica Acta, 2012, A88: 216

neral simpler and easier though numerically may not be so accurate. However, it can still offer clear and concrete physical picture which is very beneficial for our tangible feeling and understanding of the physics under study. We have to admit that quantal language like wave function is often too abstract and farther beyond our daily conceptual ability than the classical language.

This book summarizes the highlights of our work on the bond polarizability approach to the Raman intensity analysis. The topics covered include surface enhanced Raman scattering, Raman excited virtual state and ROA. The first chapter briefly introduces the Raman effect in a succinct and clear way. Chapter 2 deals with the normal mode analysis which is a basic tool for our work. Chapter 3 introduces our proposed algorithm for the Raman intensity analysis. Chapter 4 heavily introduces the physical picture of Raman virtual states. Chapter 5 presents our work that leads to a comprehensive idea about the Raman virtual states. Chapter 6 demonstrates how this bond polarizability algorithm is extended to ROA intensity analysis. Chapters 7 and 8 offer our works on ROA, showing many findings on ROA mechanism that were not known or neglected before. Chapter 9 introduces our proposed classical treatment on ROA which, as combined with the results from the bond polarizability analysis, leads to a comprehensive physical picture of the Raman process. In particular, this classical treatment unifies ROA and VCD (vibrational circular dichroism) on equal footing. In this book, in each section, there are *Comments* which summarize the key ideas and their evaluation. This will help the readers to capture the core ideas of the presentations in a logical and sequential way. In Appendix B, a summary of our publications are listed including those on phase transitions which are not covered in this book. This may be convenient to the readers, if needed.

Indeed, few monographs on Raman effect are written from the viewpoint of Raman intensity in a systematic way as this book is. I hope our works will convince the readers that many Raman and ROA features not known or neglected before can be retrieved by this intensity approach. I also hope the readers can appreciate these novel viewpoints on Raman and ROA, albeit polarizability for Raman is an old idea which has already been exploited deeply by many researchers since its discovery in 1928 and when computer simulation for Raman and ROA is becoming a daily experience nowadays.

Finally, I have to mention that these works bear the efforts of my former graduate students at the Institute of Chemistry, Chinese Academy of Sciences and Physics Department of Tsinghua University who have engaged in these studies during the past years. Also, I thank Professors P. J. Wang and H. X. Shen for their cooperation with me on the ROA work. Over the past nearly 30 years, the grant support by the China National Science Foundation and that by the State Key Laboratory of Low-dimensional Quantum Physics of the Department of Physics, Tsinghua University are greatly acknowledged.

Guozhen Wu
On the 6th of April, 2015
At Physics Department,
Tsinghua University, Beijing

Contents

Chapter 1
Raman effect

1.1 Basics: the Raman virtual state

Raman effect was discovered by C. V. Raman in 1928[1-3]. It is an inelastic two-photon process, in which the scattering photon is first absorbed by the scatterer, for instance a molecule (in the followings, we will specify the scatterer to molecule without loss of generality). The photon-perturbed molecule is excited/disturbed and then relaxes and emits a secondary photon instantaneously. What excited/disturbed in the photon-perturbed molecule is its charges/electrons(from the ground state). During this process, the energy of the excited/disturbed charges may exchange with the molecular internal quanta, such as vibration through the so-called vibronic(vibration-electronic) coupling, so that the energy of the emitted photon may be different from the scattering photon. This energy difference is called the Raman shift. Raman shift is expressed in wavenumber (cm^{-1}, wavenumber multiplied by the velocity of light, 3×10^{10} cm/sec, gives the actual frequency). Raman shift gives the information of the molecular internal motion. For molecular vibration, Raman shift gives its frequency which is related to the molecular structure (geometry and atomic masses) and bond strengths. In fact, this information can also be obtained from other effects, such as infrared absorption. Hence, we know that the Raman shift is not the core information that we can obtain from the Raman effect.

The real core information lies in the Raman intensity which offers the dynamics of the photon-perturbed molecule. The photon-perturbed molecule, in general, may not be in its excited eigenstates. In such case, we call it the non-resonant the Raman process and the photon-perturbed molecule including its relaxation, the Raman excited virtual state, or simply the Raman virtual state. However, we have to stress that the Raman virtual state is a real physical entity, not just by our imagination. The word *virtual* should not be exaggerated to imply a physical meaning. The Raman virtual state is like a wave packet.

It is a common daily experience that in a pan full of water, there can be standing waves which are shaped by the pan. When a pebble is thrown into the pan, the water inside will be splashed above or even out of the pan in a random way. Similarly, when a molecule(the pan) absorbs a light quantum(the pebble), its electronic distribution(the water)will be disturbed. The disturbed(excited)electronic distribution in such an excitation is, in general, not stationary(the splashing water) or non-resonant and does not correspond to an eigenstate (the standing wave). We know that the eigenstates are governed by the molecular nuclei and can be accurately predicted by Schroedinger equation (just as that the patterns of the standing waves are defined by the pan boundary)while the non-stationary excitation including its relaxation, called virtual state, is not well defined by the nuclei and is hard to figure out though not impossible. This is the physical concept of the Raman virtual state. Just like that we cannot say that the splashing water is not a physical reality due to it is not a standing wave, we cannot say that the Raman virtual state is not a real physical entity due to that it is simply not an eigenstate or resonant one!

Back to the core issue: how to retrieve the information of the Raman virtual state from the Raman intensity? This is the central issue for the Raman study and the core topic we will cover in this book.

Before going to the issue of Raman intensity study, we will first introduce briefly the classical and quantum mechanical treatments of the Raman effect. If the readers are familiar with them, these sections may be skipped, except the ***Comments*** in Section 1. 3.

Comments

If Raman excitation is to an eigenstate, we call the process resonant. In general, this is not the case, then the process is called nonresonant. In such case, no doubt that the electrons are disturbed and the molecule is in the so-called Raman virtual state. Hence, the virtual state is a physical reality, albeit the molecule is not in an eigenstate. Do not try to interpret the phenomenon from the vocabulary—*virtual*. It is only a word, a nomenclature for the phenomenon, not a deduction based on any physical considerations.

1.2 The classical treatment

As a molecule is hit by the visible light(called the scattering light), the nuclear motion(vibration) will not be disturbed since its frequency is much less than that of the light. However, the electrons will be disturbed by the light. This disturbance of electrons by the external light is described by the electronic polarizability, which is a measure how easily the electrons can be affected by the light. As the electrons are more tightly bound to the nuclei and harder to be disturbed by the light, the electronic polarizability is smaller, otherwise, it is larger. The dimension of the electronic polarizability is volume. Its physical significance is the space occupied by the electrons. It is also proportional to the amount of electron numbers(charges). Since the electronic distribution of a molecule(or the molecular configuration/shape) is, in general, not spherical, the electronic polarizability is direction dependent, or a tensor. However, for the convenient elucidation of its physical significance, we will not go that far but simply consider it as a scalar in this book as we will pay more attention to the electronic polarizability of a bond. This will not affect the retrieval of the physical concepts of the topics concerned. For convenience, electronic polarizability is often shortened as polarizability.

Suppose the electric field of the scattering light with angular frequency ω is

$$\varepsilon = \varepsilon_0 e^{i\omega t}$$

and the vibration is described by the normal mode(at least, in the low excitation) coordinate Q_k with frequency ω_k, as

$$Q_k = Q_k^0 \cos\omega_k t$$
$$= Q_k^0 [e^{i\omega_k t} + e^{-i\omega_k t}]/2$$

in which Q_k^0 is the coordinate at equilibrium.

The electronic polarizability α will depend on Q_k. By expanding α in terms of Q_k:

$$\alpha = \alpha^0 + \sum_k (\partial\alpha/\partial Q_k)_0 Q_k \cdots$$

then under the influence of the scattering light, the induced electric dipole μ, which is responsible for the emitted light, is

$$\mu = \alpha\varepsilon$$
$$= \left[\alpha^0 + \sum_k (\partial\alpha/\partial Q_k)_0 Q_k^0 \frac{1}{2}(e^{i\omega_k t} + e^{-i\omega_k t})\right]\varepsilon_0 e^{i\omega t}$$
$$= \alpha^0 \varepsilon_0 e^{i\omega t} + \frac{1}{2}\sum_k (\partial\alpha/\partial Q_k)_0 Q_k^0 \varepsilon_0 [e^{i(\omega+\omega_k)t} + e^{i(\omega\omega_k)t}].$$

With μ containing $e^{i\omega t}$, $e^{i(\omega+\omega_k)t}$ and $e^{i(\omega-\omega_k)t}$. This shows that the scattered light possesses frequencies $\omega \pm \omega_k$ besides ω. Its spectrum looks like that shown in Fig. 1. 1. The line with frequency ω is the Rayleigh line. Its intensity is related to α^0. The lines with frequencies $\omega \pm \omega_k$ are the Raman lines. The one with $\omega - \omega_k$ is the Stokes line and that with $\omega + \omega_k$ is the anti-Stokes line. Pictorially, the former can be considered as that the energy of one vibrational quantum is absorbed from the scattering light by the molecule while the latter as that one vibrational quantum is transferred to the scattered light from the molecule. The energy radiated by the dipole is proportional to the fourth power of its angular frequency. Hence, their intensity ratio is(no other factors will be considered for the time being):

$$\frac{I_{\text{Stokes}}}{I_{\text{antiStokes}}} = \frac{(\omega - \omega_k)^4 N_{V_k=0}}{(\omega + \omega_k)^4 N_{V_k=1}}.$$

Here, N_{V_k} is the molecular population with vibrational quantum number V_k. Though $(\omega - \omega_k)^4 < (\omega + \omega_k)^4$, the population of the ground state is much larger due to the Boltzmann distribution:

$$N_{V_k=0} \gg N_{V_k=1}$$

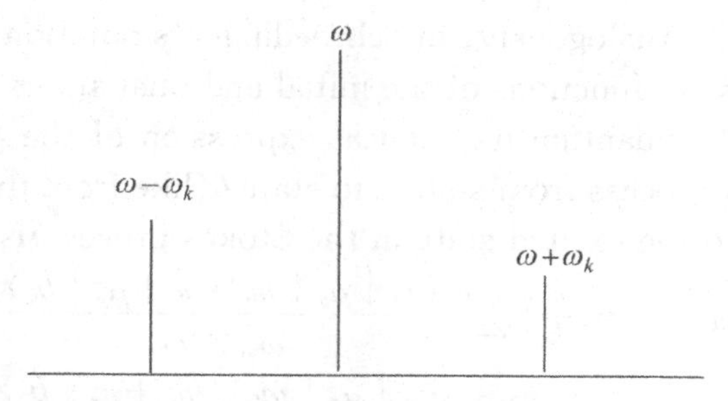

Fig. 1. 1 Rayleigh and Raman lines. Those with $\omega-\omega_k$ and $\omega+\omega_k$ are the Stokes and anti-Stokes lines, respectively.

Hence, the Stokes line is stronger than the anti-Stokes line.

α^0 is the static polarizability. It is the polarizability when the molecular motion is frozen. It shows roughly the molecular dimension. $(\partial\alpha/\partial Q_k)_0$ shows the vibronic coupling between the electronic and nuclear motions — the changing of the electronic space as the nuclei move during vibration. The Raman intensity is proportional to $(\partial\alpha/\partial Q_k)_0^2$. We stress again that the Raman intensity bears the information of the vibronic coupling which is the core physical parameter in the Raman process, while the Raman shift just does not play this core role.

Comments

Classical treatment can nicely show up the Raman shifts and that Raman is due to the vibronic coupling whose information is embedded in the intensity.

1. 3 The quantal treatment

Quantum mechanical treatment was proposed by Albrecht[4]. Here, we only offer the result of the quantal treatment, instead of its detailed derivation. In quantum mechanics, a physical process is expressed in terms of the probability of the transition from the initial state $|I\rangle$ to the final state $|F\rangle$ via H' mechanism as $\langle F|H'|I\rangle$ in Dirac's notation. (Analogously, in Schroedinger's notation, it is the integral of the wave functions of the initial and final states with the operator H'). The quantum mechanical expression of the polarizability for the Raman process from state a to state b(like from the vibrational ground state to the excited state in the Stokes process) is:

$$(\alpha_{\rho\sigma})_{ab} = \frac{1}{\hbar}\left\{\sum_{n\neq a}\frac{\langle a_0 \mid \mu_\sigma \mid n_0\rangle\langle n_0 \mid \mu_\rho \mid b_0\rangle}{\omega_{an}+\omega} + \sum_{m\neq b}\frac{\langle a_0 \mid \mu_\rho \mid m_0\rangle\langle m_0 \mid \mu_\sigma \mid b_0\rangle}{\omega_{bm}-\omega}\right\}.$$

Here, $(E_i - E_j)/\hbar = \omega_{ij}$ is the angular frequency corresponding to the energy difference between the transition states. $|n_0\rangle$ and $|m_0\rangle$ are the intermediate excited states. The Raman process is the superposition of all the paths that each starts and ends in the same initial and final states but proceeds through all possibly allowed intermediate excited states. μ_ρ and μ_σ are the polarization operators of the photons involved in the two photon process of the Raman scattering. The appearance of $\omega_{an}+\omega$ (note that ω_{an} is negative by definition) in the denominators shows that as the energy of the light quantum is very close to the energy differences of these levels, that is when $\omega_{an}+\omega\approx 0$, the scattering probability will be much larger. This is the resonance effect. Otherwise, if the energy differences of the molecular levels and the energy of the light quantum are very large, then very small the scattering probability will be. When $E_a > E_b$, the process is the anti-Stokes scattering. When $E_a < E_b$, the process is the Stokes scattering. From the viewpoint of symmetry, μ_σ and μ_ρ are of symmetry x, y or z. Hence, $\alpha_{\rho\sigma}$ is of symmetry xx, yy, zz, xy, xz, yz.

Comments

Since the Raman virtual state is not an eigenstate, in the quantal treatment, it is expressed in terms of the complete eigenstates $\{|n_i\rangle\}$ as:

$$\text{Raman virtual state} = \sum c_i \mid n_i\rangle.$$

This is shown in Fig. 1. 2, pictorially. However, we cannot deny its physical entity just because it is not an eigenstate. This expression of the Raman virtual state in terms of eigenstates is in fact quite formal since it is very hard to figure out the eigen-coefficients. Thus, we know that to elucidate the physics of the Raman virtual state from the Raman intensity, quantal method is not our first choice.

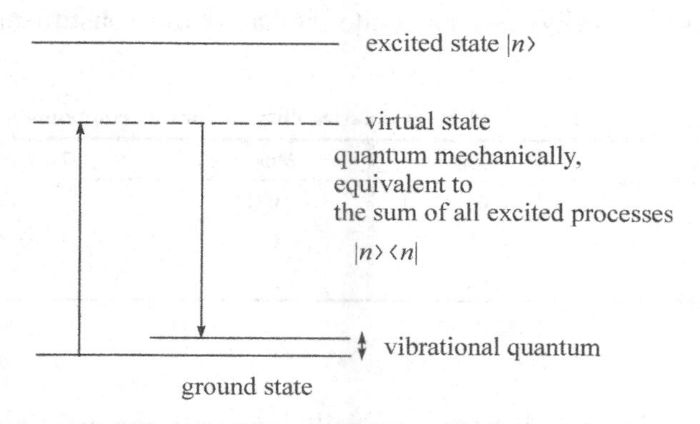

Fig. 1. 2 The Raman excited virtual state is equivalent to the sum of all the excited processes, quantum mechanically.

1. 4 Theory of bond polarizability by Wolkenstein

Wolkenstein[5] has proposed that molecular polarizability α can be written as

$$\alpha = \sum_t \alpha_t + \sum_a \alpha_a$$

in which α_t and α_a are the polarizabilities along the bonds and on the atoms, respectively.

For a mode coordinate Q_k, we have:

$$\partial\alpha/\partial Q_k = \sum_t \frac{\partial\alpha_t}{\partial S_t}\frac{\partial S_t}{\partial Q_k}\ .$$

Here, S_t is the internal coordinate. (The atomic polarizability is assumed to be of spherical shape.) Hence, in certain situations, from the Raman intensity[i. e., $(\partial\alpha/\partial Q_k)^2$]and $\partial S_t/\partial Q_k$ by the normal mode analysis(see Chapter 2), $\partial\alpha_t/\partial S_t$ can be obtained. For convenience, $\partial \alpha_t / \partial S_t$ can be called the bond polarizability. The result by Wolkenstein is shown in Table 1. 1. These C—H bonds share very similar polarizabilities though they are in different molecules. This means that molecules are built up by chemical bonds just like the compilation of bricks to form a structure. Their final shapes and properties may be very divergent, however, locally they are quite similar if the constituents are the same.

Table 1. 1 The C—H bond polarizability in various compounds.

Molecule	$\partial\alpha/\partial r_{C-H}$/Å^2	Molecule	$\partial\alpha/\partial r_{C-H}$/Å^2
CH_4	1.04	C_6H_6	1.00
C_2H_6	1.10	C_2H_2	1.02
C_2H_4	1.04		

Comments

In the following chapters, we will develop a systematic algorithm to elucidate the bond polarizabilities from Raman intensities. Through these bond polarizabilities, the nature of the Raman virtual state will be clearly demonstrated.

References

[1] Raman C V, Krishnan K S. Indian J. Phys. , 1928, 2: 387, 399.
[2] Raman C V, Krishnan K S. Nature, 1928, 121: 501, 619.
[3] Raman C V, Krishnan K S. Proc. Roy. Soc. , 1928, 122A: 23.
[4] Albrecht A C. J. Chem. Phys. , 1961, 34: 1476.
[5] Wolkenstein M, Acad C R. Sci. USSR, 1941, 32: 185.

Chapter 2
Normal mode vibration

Since we will heavily rely on the normal mode analysis in our subsequent work on the Raman intensity, we will spend this chapter on its analysis. For readers who are familiar with this topic may just skip this chapter.

2.1 Born-Oppenheimer approximation

A molecule is composed of positively charged nuclei and negatively charged electrons. Electrons are moving around nuclei. Since the electron mass is much less than the nuclear one, roughly in a ratio of 1 : 10000, the motion speed of electrons is much faster than nuclei. In terms of period, they are roughly in a ratio of 100 : 1. Therefore, in an instant of nuclear motion, the electronic distribution can keep up with the potential produced by the nuclear configuration. Meanwhile, if due to excitation like light absorption, the electronic distribution is changed, however, nuclei cannot move fast enough to the new positions where they should stay to match up with the new potential for the excited electrons. This picture concerning the electronic and nuclear motions is the foundation of Born-Oppenheimer approximation. Since the motion speed of electrons is very fast and the electron mass is very small, we can regard electrons of a molecule as a whole. This is the core concept of wave function for electrons. As to the nuclear motion, since its speed is much slower, it is called vibration. As long as there is the concept of vibration, there is the idea of vibrational mode.

Hence, we can separate molecular motions into the nuclear vibration in the potential produced by the electronic distribution and the electronic motion in the potential produced by the fixed nuclear configuration. Specifically, the nuclear motion depends only on nuclear coordinates and the electronic motion depends on both nuclear and electronic coordinates. The former is related to the normal mode analysis and the latter to the solvation of the electronic wave function.

2.2 Normal mode

The degrees of freedom of a molecule with N atoms are 3 N, in which 3 degrees are translational and the other 3 (for nonlinear molecules) or 2 (for linear molecules) degrees are rotational. Hence, the vibrational degrees of freedom are $3N-6$ or $3N-5$.

The kinetic energy can be written as

$$T=\frac{1}{2}\sum_{\alpha=1}^{N}m_\alpha\left[\left(\frac{d\Delta X_\alpha}{dt}\right)^2+\left(\frac{d\Delta Y_\alpha}{dt}\right)^2+\left(\frac{d\Delta Z_\alpha}{dt}\right)^2\right]$$

in which ΔX_α, ΔY_α and ΔZ_α are defined as

$$\Delta X_\alpha = X_\alpha-(X_\alpha)_e$$
$$\Delta Y_\alpha = Y_\alpha-(Y_\alpha)_e$$
$$\Delta Z_\alpha = Z_\alpha-(Z_\alpha)_e$$

where $(X_\alpha)_e$, $(Y_\alpha)_e$, $(Z_\alpha)_e$ are the coordinates of the atom α at equilibrium and X_α, Y_α, Z_α are its instantaneous coordinates, m_α is the mass of atom α. For convenience, we define a new set of coordinates q_i:

$$q_1=\sqrt{m_1}\Delta X_1,\ q_2=\sqrt{m_1}\Delta Y_1,\ q_3=\sqrt{m_1}\Delta Z_1,\ q_4=\sqrt{m_2}\Delta X_2,\cdots$$

then the kinetic energy becomes

$$T=\frac{1}{2}\sum_{i}^{3N}\dot{q}_i^2.$$

The potential energy V is a function of q_i,

$$V=V(q_1,\ \cdots,\ q_{3N}).$$

The expansion in terms of q_i leads to

$$V=V_0(0,\ \cdots,\ 0)+\sum_{i=1}^{3N}\left(\frac{\partial V}{\partial q_i}\right)_0 q_i+\frac{1}{2}\sum_{i=1}^{3N}\sum_{j=1}^{3N}\left(\frac{\partial^2 V}{\partial q_i\partial q_j}\right)_0 q_iq_j+\cdots.$$

At equilibrium, the potential is the least, then

$$\left(\frac{\partial V}{\partial q_i}\right)_0=0.$$

Meanwhile, we define

$$V_0 = 0$$

and keep only the second order terms (the harmonic approximation), then we have

$$V = \frac{1}{2}\sum_i \sum_j f_{ij} q_i q_j$$

in which

$$f_{ij} \equiv \left(\frac{\partial^2 V}{\partial q_i \partial q_j}\right)_0$$

is the force constant.

The kinetic energy T and the potential energy are now functions of $\dot{q}$ and q. For the Lagrangian $L = T - V$, the equations of motion are:

$$\frac{\mathrm{d}}{\mathrm{d}t}\left(\frac{\partial L}{\partial \dot{q}_i}\right) - \frac{\partial L}{\partial q_i} = 0, \quad i = 1, 2, \cdots, 3N$$

or

$$\ddot{q}_i + \sum_{j=1}^{3N} f_{ij} q_j = 0, \quad i = 1, 2, \cdots, 3N.$$

Now supposing that

$$q_i = q_i^0 \cos(\omega t + \varepsilon),$$

we have

$$\sum_{j=1}^{3N} (f_{ij} - \delta_{ij}\omega^2) q_j^0 = 0, \quad i = 1, 2, \cdots, 3N.$$

For the equation set to have nonzero solutions, it is required that:

$$\begin{vmatrix} f_{11} - \omega^2 & f_{12} & \cdots & f_{1,3N} \\ f_{21} & f_{22} - \omega^2 & \cdots & f_{2,3N} \\ \vdots & \vdots & & \vdots \\ f_{3N,1} & \cdots & \cdots & f_{3N,3N} - \omega^2 \end{vmatrix} = 0$$

or in simple notation:

$$\det | f_{ij} - \delta_{ij}\omega^2 | = 0.$$

This determinant has $3N$ solutions. (In fact, it has 6 or 5 zero solutions corresponding to translational and rotational motions.) Suppose its solutions are ω_k^2's, then the relative $q_{i,k}^0$ can be obtained. By the normalization condition:

$$\sum_i q_{i,k}^{0\ 2} = 1$$

$q_{i,k}^0$ can be determined. The normalized $q_{i,k}^0$ is expressed as L_{ik}. To each ω_k, there is a corresponding set of L_{ik}. The motion is such that all the atoms possess the same frequency ω_k, while with various amplitudes which are L_{ik}'s. Such a motion is called the normal mode.

For $\omega_k \neq 0$, the general solution is

$$q_{i,k} = \sum_{k=1}^{3N-6} C_k L_{ik} \cos(\omega_k t + \varepsilon)$$

with

$$i = 1, 2, \cdots, 3N-6(3N-5)$$

C_k's are arbitrary constants.

2.3 Normal coordinates

Besides $q_{i,k}$, normal coordinate Q_k is defined by such that the kinetic and potential energies are(this is also under the harmonic approximation)

$$T = \frac{1}{2}\sum_{k=1}^{3N-6} \dot{Q}_k^2, \quad V = \frac{1}{2}\sum_{k=1}^{3N-6} \lambda_k^2 Q_k^2$$

or in matrix notation

$$T = \frac{1}{2} P_Q^{\mathrm{T}} P_Q, \quad V = \frac{1}{2} Q^{\mathrm{T}} \Lambda Q$$

in which, elements of Λ is: $\Lambda_{ki} = \lambda_k^2 \delta_{ki}$ and

$$P_Q = \begin{bmatrix} \dot{Q}_1 \\ \dot{Q}_2 \\ \vdots \\ \dot{Q}_{3N-6} \end{bmatrix}, \quad Q = \begin{bmatrix} Q_1 \\ Q_2 \\ \vdots \\ Q_{3N-6} \end{bmatrix}.$$

There is a linear transformation between $q_{i,k}$ and Q_k:

$$q_{i,k} = \sum_{k=1}^{3N-6} l_{ik} Q_k.$$

From T, V and the Lagrangian equation, the equation for Q_k is just that of the simple harmonic oscillator:

$$\ddot{Q}_k + \lambda_k^2 Q_k = 0.$$

Suppose its solution is

$$Q_k = Q_k^0 \cos(\lambda_k t + \varepsilon).$$

Therefore, we have

$$q_{i,k} = \sum_k^{3N-6} l_{ik} Q_k^0 \cos(\lambda_k t + \varepsilon).$$

By comparing with (see Section 2. 2)

$$q_{i,k} = \sum_{k=1}^{3N-6} C_k L_{ik} \cos(\omega_k t + \varepsilon)$$

we have:

$$L_{ik} = l_{ik}, \quad \omega_k = \lambda_k.$$

This shows that the motion with Q_k as the coordinate has frequency, ω_k. Q_k and $q_{i,k}$ are then related by

$$q_{i,k} = \sum_{k=1}^{3N-6} L_{ik} Q_k$$

or in matrix notation:

$$\boldsymbol{q} = \boldsymbol{LQ}, \quad \boldsymbol{Q} = \boldsymbol{L}^{-1}\boldsymbol{q}.$$

In summary, we have the following important concepts about the molecular vibration: Under the harmonic approximation, the molecular vibration can be decomposed into a set of normal modes in which all the atoms vibrate around their equilibrium positions with phase 0 or π and with the same frequency though with different amplitudes. All the normal modes are independent. When they are expressed in normal coordinates, they are one dimensional simple harmonic oscillator.

Comments

The concept of normal modes for molecular vibration is only valid under harmonic approximation. For low vibrational excitation as in the normal Raman process, this is appropriate. However, in highly excited vibration, this concept will break down. Therein, nonlinear effect will be a prominent and unusual phenomenon like the chaos. Or on the contrary, very localized motion may appear[1]. Usual treatments for high vibration such as mode-mode coupling and the inclusion of non-linear terms, etc., may be adequate for obtaining some nonlinear parameters. However, these nonlinear effects, in nature, cannot be approximated or obtained just through such higher order expansions.

2.4 Generalized coordinates and normal mode analysis

Previously, we have defined two coordinates: q_i and Q_k. q_i is intuitive but the analysis based on it is very cumbersome. Q_k is more convenient though its configuration can only be recognized after the normal mode analysis. In reality, the realization of Q_k from q_i is not often adopted. Instead, other coordinates are adopted. For this, the concept of generalized coordinates is required.

Suppose the generalized coordinate is S_t (which can be q_i or Q_k). Let the coordinate ξ_i be:

$$\xi_1 = \Delta X_1, \ \xi_2 = \Delta Y_1, \ \xi_3 = \Delta Z_1, \ \xi_4 = \Delta X_2, \ \cdots$$

and suppose the transformation from ξ_i to S_t is

$$S_t = \sum_{i=1}^{3N-6} B_{ti}\xi_i.$$

Since the kinetic energy is

$$T = \frac{1}{2}\sum_{i=1}^{3N-6} m_i \dot{\xi}_i^2,$$

the momentum P_i that is conjugate to ξ_i is

$$P_i = \frac{\partial T}{\partial \dot{\xi}_i} = \sum_i \frac{\partial T}{\partial \dot{S}_t}\frac{\partial \dot{S}_t}{\partial \dot{\xi}_i} = \sum_t P_t B_{ti}$$

where P_t is the momentum corresponding to S_t.

The kinetic energy can be written as:

$$T = \frac{1}{2}\sum_i \frac{P_i^2}{m_i} = \frac{1}{2}\sum_i \frac{1}{m_i}\left(\sum_t P_t B_{ti}\right)^2$$

$$= \frac{1}{2}\sum_{tt'}\left(\sum_i \frac{1}{m_i} B_{ti} B_{t'i}\right) P_t P_{t'}.$$

By defining

$$G_{tt'} = \sum_i \frac{1}{m_i} B_{ti} B_{t'i},$$

we have

$$T = \frac{1}{2} \sum_{tt'} G_{tt'} P_t P_{t'}.$$

Similarly, potential V can be expanded in S_t as:

$$V = \frac{1}{2} \sum_{tt'} F_{tt'} S_t S_{t'}.$$

In matrix notation, we have:

$$T = \frac{1}{2} \boldsymbol{P}^{\mathrm{T}} \boldsymbol{G} \boldsymbol{P}$$

$$V = \frac{1}{2} \boldsymbol{S}^{\mathrm{T}} \boldsymbol{F} \boldsymbol{S}.$$

In Section 2. 3, it is known that in terms of Q and its momentum P_Q $(=\dot{Q})$, T and V have simple forms:

$$T = \frac{1}{2} \boldsymbol{P}_Q^{\mathrm{T}} \boldsymbol{P}_Q$$

$$V = \frac{1}{2} \boldsymbol{Q}^{\mathrm{T}} \boldsymbol{\Lambda} \boldsymbol{Q}.$$

If the transformation from Q to S coordinates is:

$$\boldsymbol{S} = \boldsymbol{L} \boldsymbol{Q},$$

then

$$V = \frac{1}{2} \boldsymbol{S}^{\mathrm{T}} \boldsymbol{F} \boldsymbol{S} = \frac{1}{2} \boldsymbol{Q}^{\mathrm{T}} \boldsymbol{L}^{\mathrm{T}} \boldsymbol{F} \boldsymbol{L} \boldsymbol{Q}$$

$$= \frac{1}{2} \boldsymbol{Q}^{\mathrm{T}} \boldsymbol{\Lambda} \boldsymbol{Q}.$$

Therefore,

$$\boldsymbol{\Lambda} = \boldsymbol{L}^{\mathrm{T}} \boldsymbol{F} \boldsymbol{L}.$$

On the other hand, we have

$$P_{Q_k} = \frac{\partial T}{\partial \dot{Q}_k} = \sum_t \frac{\partial T}{\partial \dot{S}_t} \frac{\partial \dot{S}_t}{\partial \dot{Q}_k} = \sum_t P_t L_{tk}$$

or in matrix notation:

$$\boldsymbol{P}_Q = \boldsymbol{L}^{\mathrm{T}} \boldsymbol{P}.$$

Combining with the expression:

$$T=\frac{1}{2}\boldsymbol{P}_Q^{\mathrm{T}}\boldsymbol{P}_Q,$$

we have

$$T=\frac{1}{2}\boldsymbol{P}^{\mathrm{T}}\boldsymbol{L}\boldsymbol{L}^{\mathrm{T}}\boldsymbol{P}.$$

Since

$$T=\frac{1}{2}\boldsymbol{P}^{\mathrm{T}}\boldsymbol{G}\boldsymbol{P},$$

then

$$\boldsymbol{G}=\boldsymbol{L}\boldsymbol{L}^{\mathrm{T}}$$

or

$$\boldsymbol{L}^{\mathrm{T}}=\boldsymbol{L}^{-1}\boldsymbol{G}.$$

Together with the following expression:

$$\boldsymbol{\Lambda}=\boldsymbol{L}^{\mathrm{T}}\boldsymbol{F}\boldsymbol{L},$$

we have:

$$\boldsymbol{\Lambda}=\boldsymbol{L}^{-1}\boldsymbol{G}\boldsymbol{F}\boldsymbol{L}.$$

Obviously, $\boldsymbol{GF}$ is diagonalized by $\boldsymbol{L}$ through a similarity transformation. This expression can be put in the form

$$\boldsymbol{G}\boldsymbol{F}\boldsymbol{L}=\boldsymbol{L}\boldsymbol{\Lambda}$$

or

$$\sum_{t'}(GF)_{tt'}L_{t'k}=L_{tk}\Lambda_{kk}$$

or

$$\sum_{t'}[(GF)_{tt'}-\delta_{tt'}\omega_k^2]L_{t'k}=0.$$

This procedure is the normal mode analysis[2]. Firstly, one has to construct $\boldsymbol{G}$ by its definition, then to obtain $\boldsymbol{F}$ matrix and finally to diagonalize $\boldsymbol{GF}$. The diagonal elements are ω_k^2's and the eigenvectors are $\boldsymbol{L}$'s which contain the contents of the normal coordinate $\boldsymbol{Q}$ (Note that $\boldsymbol{L}^{-1}\boldsymbol{S}=\boldsymbol{Q}$).

Often S_t can be chosen as the bond stretching and bending between two bonds. These are the internal coordinates.

2.5 Some useful relations among different coordinates

We summarize the relations between various coordinate systems as follows: suppose there are two coordinate systems, S_1 and S_2. (If their dimensions are not the same, then generalized coordinate transformation has to be adopted. The situation appears when redundant coordinates are chosen. See the subsequent discussion.) Their relation is

$$\boldsymbol{A}\boldsymbol{S}_1 = \boldsymbol{S}_2.$$

Then:

(1) The momenta, $\boldsymbol{P}_{S_1}$ and $\boldsymbol{P}_{S_2}$ are related by

$$\boldsymbol{A}^{\mathrm{T}}\boldsymbol{P}_{S_2} = \boldsymbol{P}_{S_1}$$

This is because $P_{S_1 i} = \dfrac{\partial T}{\partial \dot{S}_{1i}} = \sum\limits_k \dfrac{\partial T}{\partial \dot{S}_{2k}} \dfrac{\partial \dot{S}_{2k}}{\partial \dot{S}_{1i}} = \sum\limits_k P_{S_2 k} A_{ki}.$

(2) In S_1 and S_2 coordinates, T and V are

$$T_{S_1} = \frac{1}{2}\boldsymbol{P}_{S_1}^{\mathrm{T}}\boldsymbol{G}_{S_1}\boldsymbol{P}_{S_1}, \quad T_{S_2} = \frac{1}{2}\boldsymbol{P}_{S_2}^{\mathrm{T}}\boldsymbol{G}_{S_2}\boldsymbol{P}_{S_2}$$

$$V_{S_1} = \frac{1}{2}\boldsymbol{S}_1^{\mathrm{T}}\boldsymbol{F}_{S_1}\boldsymbol{S}_{S_1}, \quad V_{S_2} = \frac{1}{2}\boldsymbol{S}_2^{\mathrm{T}}\boldsymbol{F}_{S_2}\boldsymbol{S}_2$$

and

$$\boldsymbol{G}_{S_2} = \boldsymbol{A}\boldsymbol{G}_{S_1}\boldsymbol{A}^{\mathrm{T}}$$
$$\boldsymbol{F}_{S_2} = (\boldsymbol{A}^{\mathrm{T}})^{-1}\boldsymbol{F}_{S_1}\boldsymbol{A}^{-1}$$
$$\boldsymbol{G}_{S_2}\boldsymbol{F}_{S_2} = \boldsymbol{A}\boldsymbol{G}_{S_1}\boldsymbol{F}_{S_1}\boldsymbol{A}^{-1}.$$

This is because that

$$T_{S_1} = T_{S_2}$$
$$\boldsymbol{P}_{S_1}^{\mathrm{T}}\boldsymbol{G}_{S_1}\boldsymbol{P}_{S_1} = \boldsymbol{P}_{S_2}^{\mathrm{T}}\boldsymbol{G}_{S_2}\boldsymbol{P}_{S_2}.$$

Meanwhile

$$\boldsymbol{P}_{S_1} = \boldsymbol{A}^{\mathrm{T}}\boldsymbol{P}_{S_2}, \quad \boldsymbol{P}_{S_1}^{\mathrm{T}} = \boldsymbol{P}_{S_2}^{\mathrm{T}}\boldsymbol{A}.$$

Then

$$\boldsymbol{A}\boldsymbol{G}_{S_1}\boldsymbol{A}^{\mathrm{T}} = \boldsymbol{G}_{S_2}.$$

Since

$$V_{S_1} = V_{S_2}, \quad \boldsymbol{S}_1^{\mathrm{T}}\boldsymbol{F}_{S_1}\boldsymbol{S}_1 = \boldsymbol{S}_2^{\mathrm{T}}\boldsymbol{F}_{S_2}\boldsymbol{S}_2$$

together with

$$\boldsymbol{S}_2=\boldsymbol{A}\boldsymbol{S}_1$$

then

$$\boldsymbol{S}_2^{\mathrm{T}}=\boldsymbol{S}_1^{\mathrm{T}}\boldsymbol{A}^{\mathrm{T}}$$

and

$$\boldsymbol{F}_{S_2}=(\boldsymbol{A}^{\mathrm{T}})^{-1}\boldsymbol{F}_{S_1}\boldsymbol{A}^{-1}.$$

Hence,

$$\boldsymbol{G}_{S_2}\boldsymbol{F}_{S_2}=\boldsymbol{A}\boldsymbol{G}_{S_1}\boldsymbol{A}^{\mathrm{T}}(\boldsymbol{A}^{\mathrm{T}})^{-1}\boldsymbol{F}_{S_1}\boldsymbol{A}^{-1}=\boldsymbol{A}\boldsymbol{G}_{S_1}\boldsymbol{F}_{S_1}\boldsymbol{A}^{-1}.$$

(3) It can be easily deduced that

$$\boldsymbol{L}_{S_2}=\boldsymbol{A}\boldsymbol{L}_{S_1}.$$

(4) For $\boldsymbol{q}$, $\boldsymbol{Q}$, $\boldsymbol{L}^{-1}$ possesses the following property.

$$(\boldsymbol{L}^{-1})_{ki}=L_{ik}.$$

This is because if $\boldsymbol{G}_{S_1}=\boldsymbol{G}_{S_2}=\boldsymbol{I}$ for $\boldsymbol{S}_1$ and $\boldsymbol{S}_2$,
then from

$$\boldsymbol{A}\boldsymbol{G}_{S_1}\boldsymbol{A}^{\mathrm{T}}=\boldsymbol{G}_{S_2},$$

we have $\boldsymbol{A}\boldsymbol{A}^{\mathrm{T}}=\boldsymbol{I}$, i. e., $\boldsymbol{A}^{\mathrm{T}}=\boldsymbol{A}^{-1}$.

Since for $\boldsymbol{q}$ and $\boldsymbol{Q}$, their $\boldsymbol{G}=\boldsymbol{I}$. Hence, $(\boldsymbol{L}^{-1})_{ki}=L_{ik}$.

(5) For a molecule of N atoms, internal coordinates $\boldsymbol{R}$ of bond stretches and bendings can be adopted. In such a choice, there are often redundant coordinates, i. e., the number of internal coordinates chosen is larger than that of normal modes, $3N-6$. To delete the redundant coordinates, one may adopt a set of $3N-6$ independent and irredundant coordinates according to the structure and symmetry of the molecule. These coordinates are often the symmetry coordinates.

Suppose the number of internal coordinates $\boldsymbol{R}$ is m which is larger than $3N-6$. Their relation with symmetry coordinates $\boldsymbol{S}$ is $\boldsymbol{S}=\boldsymbol{B}_S\boldsymbol{R}$. Furthermore, the transformation between the force fields $\boldsymbol{F}_R$ and $\boldsymbol{F}_S$ in these two coordinates is $\boldsymbol{F}_R=\boldsymbol{B}_S^{\mathrm{T}}\boldsymbol{F}_S\boldsymbol{B}_S$. Since $\boldsymbol{B}_S$ is of dimension $3N-6\times m$, we have to adopt the following transformation in order to deduce $\boldsymbol{F}_S$ from $\boldsymbol{F}_R$.

By $\boldsymbol{F}_R=\boldsymbol{B}_S^{\mathrm{T}}\boldsymbol{F}_S\boldsymbol{B}_S$, we have $\boldsymbol{B}_S\boldsymbol{F}_R\boldsymbol{B}_S^{\mathrm{T}}=(\boldsymbol{B}_S\boldsymbol{B}_S^{\mathrm{T}})\boldsymbol{F}_S(\boldsymbol{B}_S\boldsymbol{B}_S^{\mathrm{T}})$. Now, $\boldsymbol{B}_S\boldsymbol{B}_S^{\mathrm{T}}$ is a symmetric square matrix of dimension $3N-6$ and its inverse is also symmetric. We then have,

$$F_S = (B_S B_S^T)^{-1} B_S F_R B_S^T (B_S B_S^T)^{-1}$$

or

$$F_S = [B_S^T (B_S B_S^T)^{-1}]^T F_R [B_S^T (B_S B_S^T)^{-1}].$$

This enables us to deduce F_S from F_R.

Comments

Usually, normal mode analysis can be taken the following procedures. The soft wares for these procedures are common.

(1) Optimizing the molecule structure by Gaussian software package, which also outputs the atomic Cartesian coordinates and the corresponding force fields.

(2) Construct the symmetry coordinates from the internal coordinates and obtain G_S and F_S under the symmetry coordinates.

(3) Diagonalize $G_S F_S$ and output mode eigenfrequencies ω_k, eigenvectors L_{ik} and the potential energy distribution (PED), which is defined as $L_{ik}^2 F_{ii} / \omega_k^2$ for coordinate i in mode k. PED shows the contribution of a local coordinate to a normal mode.

(4) If needed, refine the force fields, so as to fit the calculated mode frequencies to the experimental ones.

References

[1] Wu G. Nonlinearity and Chaos in Molecular Vibrations. Elsevier, Amsterdam, 2005.
[2] Wilson Jr. E B, Decius J C, Cross P C. Molecular Vibrations. New York: McGraw-Hill, 1955.

Chapter 3

The elucidation of bond polarizabilities

3.1 Raman intensity in the temporal domain

We generally refer the Raman intensity for a vibrational mode to the area under the spectral contour in the frequency domain. If the spectral intensity is in the frequency domain, $I'_j(\nu)$, then it may be Fourier transformed to the temporal domain, $I_j(t)$. These two intensities are related by $\int I'_j(\nu)e^{i2\pi\nu t}d\nu = I_j(t)$. We note that for $t=0$, this transformation reduces to $\int I'_j(\nu)d\nu = I_j(t=0) \equiv I_{j0}$. Hence, the integrated Raman intensity over ν is just the temporal intensity at $t=0$.

3.2 The elucidation of bond polarizabilities

From Chapter 1, we know that the Raman intensity is proportional to the fourth power of frequency. For the Raman process, the electronic polarizability can be expressed in terms of $(\partial\alpha/\partial Q_j)Q_j$ for mode j (the subscript 0 in $(\partial\alpha/\partial Q_j)_0$ as adopted in Chapter 1 will not be shown for short, hereafter.). The Raman intensity in the frequency domain is $I_{j0}\approx(\nu_0-\nu_j)^4(\partial\alpha/\partial Q_j)^2Q_j^2$. Here ν_0 and ν_j are frequencies of the exciting laser and that corresponding to the Raman shift. They can be in cm^{-1}. We note that

$$I_{j0}\approx(\nu_0-\nu_j)^4(\partial\alpha/\partial Q_j)^2Q_j$$

is

$$I_j(t=0)\approx(\nu_0-\nu_j)^4(\partial\alpha/\partial Q_j)^2Q_j^2 \quad (t=0).$$

We may extrapolate it to time t as

$$I_j(t)\approx(\nu_0-\nu_j)^4(\partial\alpha/\partial Q_j)^2Q_j^2(t).$$

Since $I_j(t)$ is in the range of 1 to 10 ps and $Q_j(t)$ is in the range of 0.1 to 0.01 ps, we leave $Q_j(t)$ to the Raman transition moment. Then, $\langle n_f | Q_j | n_i \rangle \approx 1/\sqrt{\nu_j}$, with $|n_i\rangle$ and $|n_f\rangle$ the initial and final vibrational states in the Raman process and we have

$$I_j(t) \approx \frac{(\nu_0 - \nu_j)^4}{\nu_j} (\partial\alpha(t)/\partial Q_j)^2.$$

This is a temporal extension of the formula given by Chantry[1]:

$$I_{j0} \approx (\nu_0 - \nu_j)^4 / \nu_j \ (\partial\alpha/\partial Q_j)^2.$$

This treatment is semi-classical or the quantum analogue of the classical treatment.

From the Raman intensity, it is easy to obtain $(\partial\alpha/\partial Q_j)$. However, $(\partial\alpha/\partial Q_j)$ does not offer us much information concerning the details of the Raman process, i.e., the electronic information of the Raman virtual state since the normal coordinate Q_j is a *global* one. Then, the transformation from $(\partial\alpha/\partial Q_j)$ to $(\partial\alpha/\partial R_t)$ with R_t the bond (or symmetry) coordinate is a crucial step. For brevity, $\partial\alpha/\partial R_t$ will be called the bond (stretch/bend) polarizability. It is an indication of the distributed charges throughout the entire molecule as the result of the motion of an individual bond (stretch/bend) coordinate. However, much of the redistribution of charges is on or close to the bond. In this sense, bond stretch polarizability is an indication of the disturbed charge density on the individual bond during the Raman process. Of course, this information is mapped out through the vibronic coupling. We can understand that much information of the Raman virtual state can be obtained from the bond polarizability, especially when it is compared with the bond density of the electronic ground state.

By transforming Q_j's to the internal coordinates (or symmetry coordinates) R_t's through

$$R_t = \sum L_{tj} Q_j$$

which can be obtained from the normal mode analysis (see Chapter 2), we have:

$$\pm\sqrt{I_j(t)} \approx (\nu_0 - \nu)^2 / \sqrt{\nu_j} \sum L_{tj} [\partial\alpha(t)/\partial R_t].$$

In matrix notation, if only relative intensities are considered, we have

$$\begin{bmatrix} \partial\alpha(t)/\partial R_1 \\ \partial\alpha(t)/\partial R_2 \\ \vdots \\ \partial\alpha(t)/\partial R_{3N-6} \end{bmatrix} = [a_{jt}]^{-1} \begin{bmatrix} P_1\sqrt{I_1(t)} \\ P_2\sqrt{I_2(t)} \\ \vdots \\ P_{3N-6}\sqrt{I_{3N-6}(t)} \end{bmatrix}$$

where $a_{jt} = (\nu_0 - \nu_j)^2/\sqrt{\nu_j}L_{tj}$ and P_j is $+1$ or -1.

From the above derivation, it is clear that $(\partial\alpha(t)/\partial R_t)$ is obtainable if $[a_{jt}]^{-1}$ matrix, $I_j(t)$'s and a $\{P_j\}$ set are given. $[a_{jt}]^{-1}$ can be known from the normal mode analysis and $I_j(t)$'s could be obtained from the Fourier transform of the Raman profiles (say, in cm^{-1}) by the experiment.

The determination of $\{P_j\}$, the so-called phase problem, is crucial in this algorithm. The reason that we have this phase problem is simply that since $I_j \approx [\sum L_{tj}(\partial\alpha/\partial R_t)]^2$, from the intensity, we can only determine $|\sum L_{tj}(\partial\alpha/\partial R_t)|$ (the absolute value) but not its sign (phase). Hence, we just cannot simply determine $(\partial\alpha/\partial R_t)$ without phases known from intensities. This is quite analogous to the X-ray structure determination in which, one not only measures the diffracted spot intensities, but also has to nail down their phases in order to solve the structure. Luckily, here, we only have two phases, either $+$ or $-$ to determine.

More explicitly, suppose all the stretching and bending motions are independent, i. e., there are no couplings among the bond coordinates, then we will have $3N-6$ Raman intensities (in general) with each corresponding to a $(\partial\alpha/\partial R_t)$. Then, we have

$$\partial\alpha/\partial R_t \approx \pm\sqrt{I_t}\left[\sqrt{\nu_t}/(\nu_0-\nu_t)^2\right].$$

Note that the sign of $(\partial\alpha/\partial R_t)$ is indeterminate. Of course, molecular vibrations cannot be so simple that for a mode intensity, several $(\partial\alpha/\partial R_t)$'s will contribute to it. Our algorithm proposed is just the way to unzip all $(\partial\alpha/\partial R_t)$'s which are hidden behind the mode intensities.

Comments

(1) The relationship between the bond polarizabilities and the

square root of Raman intensities is a linear transformation. In this sense, they are equivalent. From the experiment, we observe Raman intensities. The bond polarizabilities offer us another angle to view them. However, there is the difference. Bond polarizabilities are the *remnants* of the intensities after excluding those quantities that are irrelevant to the Raman process, i. e. , the vibrational frequencies and amplitudes, L_{tj}, which are related to the atomic masses, force constants and molecular geometry. Hence, bond polarizabilities bear the core information of Raman process(not the vibrational frequencies and amplitudes, or the Raman shifts).

(2) This algorithm enables us a systematic way to explore bond polarizabilities from the Raman intensities. This method relies on a set of mode intensities, instead of individual ones. Hence, even for the case where certain intensities are not accurate, as long as they are minor, their impact on the overall elucidated bond polarizabilities may not be fatal. This is one merit of this algorithm.

(3) Sometimes, people are more familiar with the correlation treatment. For this, we know that, at $t>0$, $I_j(t)$ is related to the product(correlation) of the polarizability derivative at time 0 and time t, i. e. ,

$$I_j(t) \approx [\partial\alpha(t)/\partial Q_j][\partial\alpha(0)/\partial Q_j].$$

By supposing

$$\partial\alpha(t)/\partial Q_j = [\partial\alpha(0)/\partial Q_j] f(t)$$

where $f(t)$ is a decaying function, then

$$I_j(t) \approx [\partial\alpha(0)/\partial Q_j]^2 f(t) = [(\partial\alpha(0)/\partial Q_j)\sqrt{f(t)}]^2.$$

We may regard

$$[\partial\alpha(0)/\partial Q_j]\sqrt{f(t)}$$

as the *formal* polarizability derivative and write it as $[\partial a(t)/\partial Q_j]$, then

$$I_j(t) \approx [\partial a(t)/\partial Q_j]^2$$

Note that

$$[\partial a(0)/\partial Q_j] = [\partial\alpha(0)/\partial Q_j] \text{ since } f(0)=1.$$

Hence, we can relate $I_j(t)$ to the square of the (formal) polarizability derivative at time t. For simplicity and certainty, we may make no difference between $(\partial a(t)/\partial Q_j)$ and $(\partial \alpha(t)/\partial Q_j)$.

If $f(t)$ is an exponential function, then the decaying characteristic times of $(\partial a(t)/\partial Q_j)$ and $(\partial \alpha(t)/\partial Q_j)$ differ by a factor of 2. In this book, we refer to that of the *formal* polarizability derivative.

From $I_j(t) \approx [(\partial \alpha(0)/\partial Q_j)]^2 f(t)$ and $I_j(0) \approx [(\partial \alpha(0)/\partial Q_j)]^2$, we have $I_j(t) \approx I_j(0) f(t)$. $I_j(t)$ can be obtained from the Fourier transform of $I'_j(\nu)$ in the wavenumber domain by re-setting its central wavenumber to 0.

3.3 The phase problem

For the phase determination, in principle, various sets of $\{P_j\}$ can be tried to obtain $(\partial \alpha/\partial R_t)$ which are then checked with physical considerations to rule out the inadequate $\{P_j\}$ sets. The criterion, for instance, can be that for the same kind bond stretchings, their signs of $(\partial \alpha/\partial R_t)$ are the same and their magnitudes are of the same order. Furthermore, we may require the bond stretching polarizabilities, like C—C and C—H, to be positive. This is plausible since as a bond is stretched out, the charges on the bond will be less bound and its polarizability is larger. We note that the sets $\{P_1, P_2, \cdots\}$ and $\{-P_1, -P_2, \cdots\}$ correspond to an identical bond polarizability set just of opposite signs. This reduces the number of $\{P_j\}$ sets for checking. In our numerous works, we often find that very few $\{P_j\}$ sets survive, even that there is but one set left. For the case of multiple sets left, often their bond polarizabilities are quite consistent. This is understandable since for those I_j's which are not significantly large, their phases will not affect the elucidated bond polarizabilities significantly. This occurs frequently for those bending I_j's. Also, in our study, it never happens that there are no solutions. All these convince us that our criteria for the phase determination are physically adequate.

Comments

If one has experience with the X-ray structure determination, he would know that there is the phase problem. Once the phases are solved, the structure is solved at hand. The phase issue in Raman is analogous to that in the X-ray structure determination since both are the scattering phenomena, albeit in different energy realms and involving different physical processes. In the X-ray structure determination, the electron density that is to be solved from the diffracted intensity is analogous to the bond polarizability in our formulation.

3.4 More on the coordinate choices

For a molecular system possessing a certain set of symmetry elements, its coordinates can be grouped into classes of various symmetries. Then, $[a_{jt}]$ matrix can be block diagonalized according to the coordinate classes. For instance, for a system with S_i^A, S_i^B coordinate classes of A, B symmetries, respectively, we have

$$\begin{bmatrix} P_1^A I_1^A \\ \vdots \\ P_1^B I_1^B \\ \vdots \end{bmatrix} = \left[\begin{array}{c|c} a_{ij}^A & 0 \\ \hline 0 & a_{ij}^B \end{array}\right] \begin{bmatrix} \partial\alpha^A/\partial S_1^A \\ \vdots \\ \partial\alpha^B/\partial S_1^B \\ \vdots \end{bmatrix}.$$

For the A class, we have those coordinates from stretching and bending, respectively, $\{S_1^A, \cdots, S_m^A, S_{m+1}^A \cdots, S_r^A\}$, where $\{S_1^A, \cdots, S_m^A\}$ are the stretching coordinates and $\{S_{m+1}^A, \cdots, S_r^A\}$ are the bending coordinates. The matrix equation looks like:

$$\begin{bmatrix} P_1^A I_1^A \\ \vdots \\ P_r^A I_r^A \end{bmatrix} = \overset{\displaystyle S_1^A \ \cdots \ S_m^A \ \cdots \ S_r^A}{\begin{bmatrix} & & \\ & a_{ij}^A & \\ & & \end{bmatrix}} \begin{bmatrix} \partial\alpha^A/\partial S_1^A \\ \vdots \\ \partial\alpha^A/\partial S_m^A \\ \vdots \\ \partial\alpha^A/\partial S_r^A \end{bmatrix}.$$

If only the bond stretching part is considered, i. e. , its coupling with the bending coordinates is neglected, being called the Wolkenstein approximation, then we have:

$$\begin{bmatrix} P_1^A I_1^A \\ \vdots \\ P_m^A I_m^A \end{bmatrix} = \overset{\begin{matrix} S_1^A & \cdots & S_m^A \end{matrix}}{\begin{bmatrix} \cdot & \cdot & \cdot & \cdot & \cdot \\ \cdot & \cdot & \cdot & \cdot & \cdot \\ \cdot & \cdot & a_{ij}^A & \cdot & \cdot \\ \cdot & \cdot & \cdot & \cdot & \cdot \\ \cdot & \cdot & \cdot & \cdot & \cdot \end{bmatrix}} \begin{bmatrix} \partial\alpha^A/\partial S_1^A \\ \vdots \\ \partial\alpha^A/\partial S_m^A \end{bmatrix}.$$

Here, I_1^A, ⋯, I_m^A have to be those whose coordinates are mostly due to $\{S_1^A, \cdots, S_m^A\}$.

Back to the previous matrix equation,

$$\begin{bmatrix} \partial\alpha^A/\partial S_1^A \\ \vdots \\ \partial\alpha^A/\partial S_m^A \\ \vdots \\ \partial\alpha^A/\partial S_r^A \end{bmatrix} = \begin{bmatrix} & & \\ & a_{ij}^A & \\ & & \end{bmatrix}^{-1} \begin{bmatrix} P_1^A I_1^A \\ \vdots \\ P_r^A I_r^A \end{bmatrix}$$

in determining $\{P_1^A, \cdots, P_r^A\}$, we may only pay attention to the elucidated $\{\partial\alpha^A/S_1^A, \cdots, \partial\alpha^A/S_m^A\}$, i. e. , the stretching part. Often, for such a situation, the solutions are usually multiple. One reason is that different phases from bending intensities may affect the bond stretching polarizabilities, more or less, though not too much, due to that the coupling between stretching and bending motions is usually not too strong. We note that only when stretching and bending motions are completely independent, $\{I_1^A, \cdots, I_r^A\}$ can be grouped into these two motions separately, otherwise, they always contain both motions.

As soon as $\partial\alpha^A/S_i^A$ is obtained, $(\partial\alpha^A/\partial R_t)$ can be obtained readily. For instance, if we have $S^A=(R_1+R_2)/\sqrt{2}$, since $\partial\alpha^A/\partial R_1=\partial\alpha^A/\partial R_2$, then $\partial\alpha^A/\partial S^A=\sqrt{2}\,\partial\alpha^A/\partial R_1$. For $\partial\alpha^B/S_i^B$, since $S^B=(R_1-R_2)/\sqrt{2}$, then $\partial\alpha^B/\partial S^B=\sqrt{2}\,\partial\alpha^B/\partial R_1$, by assuming $\partial\alpha^B/\partial R_1=-\partial\alpha^B/\partial R_2$. In general, the intensity due to anti-symmetric mode is less, its polarizability is also smaller.

The restriction of the matrix equation for bond polarizabilities to the stretching part by tossing away the bending motion, of course, affects the accuracy of the obtained stretching polarizabilities. However, in many cases, due to their weak coupling, the inaccuracy caused is still tolerable. What deserves attention is that this restriction often does not hurt the physics we would draw from the bond polarizability analysis, since the physics mainly lies on the stretching polarizabilities, which is larger than the bending polarizabilities, in general. Hence, this approximation is still worthwhile, especially, when only the physical picture and not the physical quantity is of our main concern. This restriction not only reduces the dimension of the matrix equation but more important, is that it reduces our effort in solving the phase problem.

Then the issue accompanied is how to choose those intensities whose components are more of the stretching coordinates. This can be judged by the potential energy distribution (PED) for each mode, which is the output in the normal mode analysis. Sometimes, we find that a mode contains quite significant bending coordinates in addition to prominent stretching coordinates. Under such situation, will the tossing away of the bending coordinates cause significant inaccuracy in the elucidation of the stretching polarizabilities? The situation may not be so bad. Though the $\boldsymbol{L}$ elements, and hence PED for these bending coordinates are significant, the corresponding bending polarizabilities can still be small if the intensities due to the bending coordinates are small. Under such situation, the effect by the deletion of these bending coordinates can still be small. That is, the contribution to the intensity is mainly from the stretching coordinates and not from the bending coordinates. For this issue, one may refer to Appendix A.

Comments

If one is familiar with the group representation theory, he would understand the factorization of the $[a_{jt}]$ matrix into diagonal submatrices as the symmetry coordinates (coordinates that are the bases of the irreducible representations of the molecular symmetry group, or that belong to specific symmetries) are adopted. Knowledge of group representation theory is fundamental in molecular spectroscopy, though we will not cover it in this book.

3.5 The intensity measurement

We have to stress that only relative intensities and hence relative bond polarizabilities are of our concern. This greatly reduces the inaccuracies due to the instrumentation. These relative quantities are enough for our quest of the physics of the Raman topics that we are concerned with. For more accurate measurement of the Raman intensity, the effects by the grating mirror due to its differential reflectivities in different frequency domains, along with various polarizations and by the detector due to its differential responses in different frequency domains, have to be considered and corrected, if needed. For the modern state-of-the-art instrumentation, these corrections are becoming unnecessary.

3.6 The comparison to the bond electronic density

Bond polarizability is a measure of the charge distribution of the Raman virtual state. It is helpful to compare it to the bond electronic densities ρ_t of the ground state which, for instance, can be calculated by the density functional theory (like DFT with ub3lyp/cc-pvDZ). In fact, even simpler EHMO calculation can be helpful in certain cases. Sometimes, we need to deal with the highest occupied molecular orbital (HOMO) or lowest unoccupied molecular orbital (LUMO) (The reason is to be addressed in the following chapters.). Then, we may employ RHF/6-311G for the calculation, for instance. The bond electronic density is

$$\rho_t = \sum_{\gamma} \sum_{ij} C_{i\gamma}^{\alpha} C_{j\gamma}^{\beta} S_{ij}^{\alpha\beta}$$

where i, j are the labels, respectively, for the atomic orbitals of atoms α and β which form the bond t. γ is the labeling for the occupied molecular orbitals. C and S are the linear combination coefficient and the overlapping integral, respectively.

Sometimes, as we plot $(\partial\alpha(t)/\partial R_t)$ against t, we find that it can be fitted by $A\exp(-t/t_c)$. This allows us to obtain the characteristic time parameter t_c.

We will find in the subsequent works that with these parameters, much information of Raman systems can be obtained.

These conclude our preparations for the elucidation of the bond polarizabilities from Raman intensities

References

[1] Chantry G W. in The Raman Effect, Vol. 1 A. Anderson(Ed.). New York: Marcel Dekker Inc. , 1971.

Chapter 4

The Raman virtual states

4.1 The case of 2-aminopyridine

Fig. 4. 1 shows the structure and atomic numbering of the 2-aminopyridine molecule. Fig. 4. 2 shows its Raman spectra by 632. 8 nm and 514. 5 nm excitations. No significant variation of the peak positions is observed by these two excitations. However, there are variations in their intensities as shown in Table 4. 1. It shows the experimental and fitted Raman shifts, together with their potential energy distributions and relative intensities by 514. 5 nm and 632. 8 nm excitations. During the normal mode analysis, initial force constants were obtained by DFT(ub3lyp/cc-pvDZ)and then refined to fit the experimental wavenumbers. There are 13 bonds in 2-aminopyridine molecule. Hence, there are 13 modes, as marked by * in Fig. 4. 2, that possess larger portion in bond stretch motion and are employed for the elucidation of the bond polarizabilities. Fig. 4. 3 shows the plots of the bond polarizabilities under 514. 5 nm and 632. 8 nm excitations together with the bond electronic densities of the ground state calculated by RHF/6-31G*.

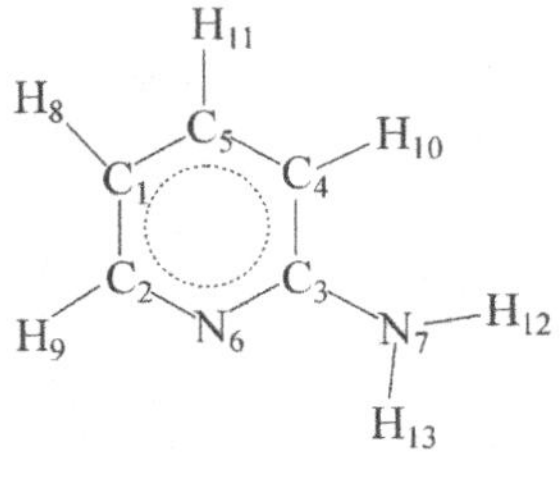

Fig. 4. 1 The atomic numbering of the 2-aminopyridine molecule.

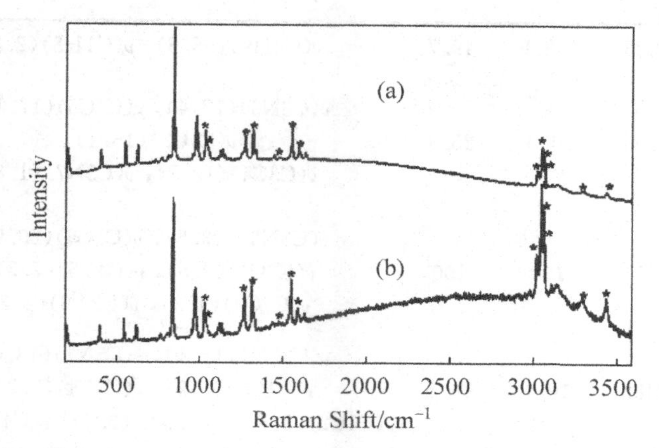

Fig. 4. 2 The Raman spectra of 2-aminopyrimidine by(a)632. 8 nm and(b)5145 nm excitations. No significant variation of the peak positions is observed by these two excitations. However, there are variations in their intensities as shown in Table 1. * shows those modes whose intensities are employed for the elucidation of the bond stretch polarizabilities.

Table 4. 1 The experimental and fitted Raman shifts, together with their potential energy distributions and relative intensities by 514. 5 nm and 632. 8 nm excitations. The intensities at 1558 cm^{-1} by both excitations are normalized to 100 for convenience. Only those distributions in stretches are listed. These are the 13 modes that possess larger portion in bond stretch motion and are employed for the elucidation of the bond polarizabilities.

Raman Shift/cm^{-1}		Intensity		Potential Energy Distribution
Exp.	Fitted	514. 5nm	632. 8nm	
3447	3433	109. 4	33. 4	ν(N7H12, N7H13)$_{as}$(100)
3303	3327	79. 4	30. 0	ν(N7H12, N7H13)$_s$(100)
3076	3079	125. 3	51. 8	ν(C1H8)(84. 8), ν(C5H11)(9. 8), ν(C2H9)(2. 3), ν(C4H10)(2. 2)
3060	3060	300. 7	146. 9	ν(C5H11)(89. 2), ν(C1H8)(9. 2)
3045	3045	41. 1	18. 8	ν(C4H10)(95. 8), ν(C1H8)(2. 2), ν(C2H9)(1. 1)

3031	3025	122.6	49.7	ν(C2H9)(95.3), ν(C1H8)(2.2)
1599	1593	24.4	25.4	ν(C2N6)(17.4), ν(C4C5)(17.1), ν(C1C2)(16.1), ν(C3C4)(12.7), ν(C3N7)(1.8)
1558	1571	100	100	ν(C3N6)(23.5), ν(C1C5)(20.9), ν(C3C4)(9.5), ν(C4C5)(7.3), ν(C1C2)(4.3), ν(C2N6)(2.2)
1483	1485	10.1	14.5	ν(C3N6)(14.4), ν(C3N7)(13.6), ν(C1C5)(5.2), ν(C3C4)(5.0), ν(C1C2)(3.8), ν(C2N6)(3.4), ν(C4C5)(1.6)
1325	1319	76.1	73.9	ν(C3N7)(41.8), ν(C2N6)(17.6), ν(C4C5)(3.5), ν(C1C5)(2.0), ν(C3C4)(1.0)
1280	1285	102.8	86.6	ν(C4C5)(16.2), ν(C2N6)(16.2), ν(C3N6)(15.7), ν(C1C2)(14.7), ν(C1C5)(6.8), ν(C3N7)(2.2)
1051	1051	80.9	101.1	ν(C1C2)(16.1), ν(C3N6)(12.7), ν(C3C4)(8.4), ν(C2N6)(5.3), ν(C1C5)(5.1), ν(C4C5)(4.7)
1042	1026	59.5	73.1	ν(C1C5)(40.9), ν(C1C2)(21.5), ν(C4C5)(6.9), ν(C3C4)(3.0), ν(C3N6)(1.8)

From Fig. 4.3, we note that

(1) The bond electronic densities of C—C bonds are the largest. Those of C—N, C—H and N—H bonds decrease in sequence, except that C3—N7 is extraordinarily small. However, the bond polarizabilities of C—N and C—C bonds are comparable. Especially, those of C4—C5, C1—C5 and C1—C2 are much depressed (with respect to C3—C4) and those of C—N bonds (C3—N6, C2—N6), especially C3—N7 bond, are enhanced greatly. This implies that the charges on C—C bonds (most possibly, this is also true for C3—C4) are flown away and toward C—N bonds, especially, C3—N7 bond in the Raman

process. We also note that the bond polarizabilities of C—H(especially, that of C5—H11)are comparable to those of C4—C5, C1—C2 and C1—C5 bonds and those of N—H are close to those of C—H or even greater(than those of C4—H10 and C1—H8). All these hint strongly that charges of the virtual state do spread to the outer space of the molecule, i. e. ,C—H and N—H bonds.

(2) The differentiation by 632. 8 nm and 514. 5 nm excitations is also significant in that by 514. 5 nm, the bond polarizabilities of C—H and N—H enhance more significantly. Apparently, by the more energetic 514. 5 nm excitation, in the virtual state, charges are flown more out of the heterocyclic ring toward the peripheral C—H and N—H bonds. If one is careful enough in looking at Table 1, this result is, in fact, demonstrated in their respective intensities by 514. 5 nm and 632. 8 nm excitations. As shown therein, all mode intensities relating to C—H and N—H stretches are more enhanced by the energetic 514. 5 nm excitation as mode intensities at 1558 cm^{-1} are normalized to the same value for this comparison.

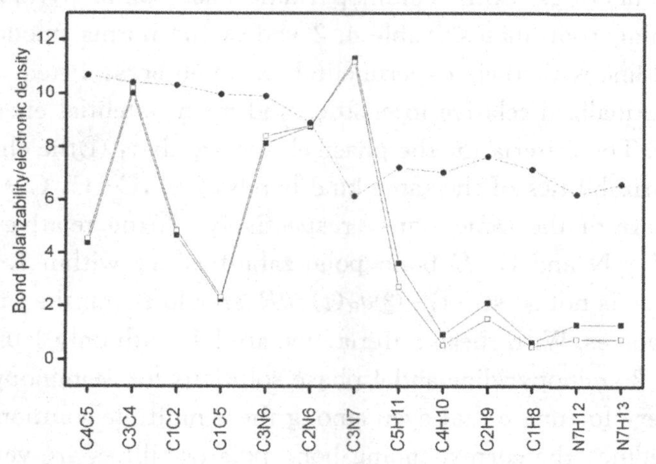

Fig. 4. 3 The plots of the bond polarizabilities under 514. 5 nm and 632. 8 nm excitations together with the bond electronic densities calculated by RHF/6-31G*. For the convenience of inference, the value of C3—C4 bond is normalized to 10. The bond polarizabilities: (■)for 514. 5 nm, (□)for 632. 8 nm; (●)for bond electronic densities.

Comments

The results we obtained from the case of 2-amino pyridine are quite general about the Raman virtual state, i. e. , the disturbed or excited charges tend to flow to the molecular periphery. This is quite natural due to the electronic repulsion among the electrons. We know that the Raman peak of C—H bond is generally rather strong despite of its small amount of charges in the ground state. We then conjecture that during the Raman process, charges will aggregate more on those peripheral C—H bonds, resulting in their strong intensities.

4. 2 More with the case of 3-aminopyridine

The case of 3-aminopyridine is appropriate for the comparison with 2-aminopyridine. Fig. 4. 4 shows the Raman spectra of 2-aminopyridine and 3-aminopyridine under 514. 5 nm excitation. Fig. 4. 5 shows the formal structures of 2-aminopyridine and 3-aminopyridine and their atomic numberings. Both 3-aminopyridine and 2-aminopyridine have 13 stretching coordinates. Table 4. 2 shows the normal modes of 3-aminopyridine with their experimental wavenumbers, fitted wavenumbers, normalized relative intensities and main potential energy distributions. The criteria for the phase choice are that: ①the elucidated bond polarizabilities of the same kind bonds, i. e. ,C—C, C—N, C—H bonds are of the same signs, respectively; ② the relative magnitudes of C—N and C—C bond polarizabilities are within 1. 5 fold. (This value is not so strict); ③$\partial\alpha(t)/\partial R_t$ should retain the same sign as time evolves. With these criteria, we are left with only 1 phase solution for 2-aminopyridine and 4 phase solutions for 3-aminopyridine. We are very fortunate that even among these multiple solutions for 3-aminopyridine, the corresponding bond polarizabilities are very similar. Thus, the most crucial step in our algorithm is solved. The temporal bond polarizabilities of 2- and 3-aminopridine under 514. 5 nm excitation are shown in Fig. 4. 6. The bond polarizabilities at $t=0$ and

5 ps near the final stage of relaxation are shown in Fig. 4. 7. Also attached are the bond electronic densities of the ground state calculated by all the occupied MO's and by only the upper eight occupied MO's. (Why this is significant will be clear in the latter discussion.)

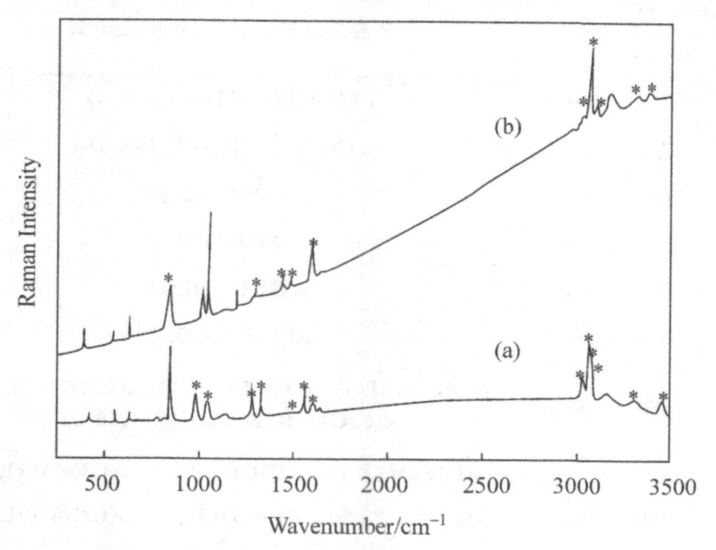

Fig. 4. 4 The Raman spectra of (a) 2-aminopyridine and (b) 3-aminopyridine under 514. 5nm excitation. * shows the peaks that appear in Tables 4. 1 and 4. 2 and are employed for the elucidation of the bond polarizabilities.

(a) (b) (c)

Fig. 4. 5 The formal structures of (a) 2-aminopyridine and (b) 3-aminopyridine and their atomic numberings. The thicker sticks show the bonds that possess the most significant polarizabilities at the very initial moment of the Raman excitation. (c) shows the bond electronic densities of the ground state.

Table 4.2 The normal modes of 3-aminopyridine with their experimental wavenumbers, fitted wavenumbers, normalized relative intensities and main potential energy distributions. For simplicity, only those of the stretching coordinates are listed and are denoted by ν.

Wavenumber/cm^{-1}		Intensity	Potential Energy Distribution/%
Expt'l.	Fitted		
3381	3371	20.2	ν(N7H12, N7H13)$_{as}$(99.4)
3313	3325	25.5	ν(N7H12, N7H13)$_{s}$(98.3)
3096	3100	14.6	ν(C1H8)(95.3)
3063	3065	100.0	ν(C5H11)(96.6)
3037	3027	16.6	ν(C2H9)(97.4)
—	3018	—	ν(C3H10)(95.2)
1590	1580	86.3	ν(C1C5)(31.6), ν(C4C5)(29.2), ν(C3N7)(12.2), ν(C3C4)(10.8), ν(C1C2)(10.0)
—	1562	—	ν(C1C2)(33.7), ν(C3C4)(29.7), ν(C3N7)(11.7)
1489	1479	40.2	ν(C4C5)(23.5), ν(C1C5)(19.1), ν(C2N6)(12.1)
1439	1445	47.6	ν(C1C2)(24.5), ν(C3C4)(23.1)
—	1354	—	ν(C3N6)(25.4), ν(C2N6)(22.8), ν(C1C5)(11.3)
1293	1299	60.9	ν(C4N7)(66.8)
842	835	91.2	ν(C4N7)(59.8)

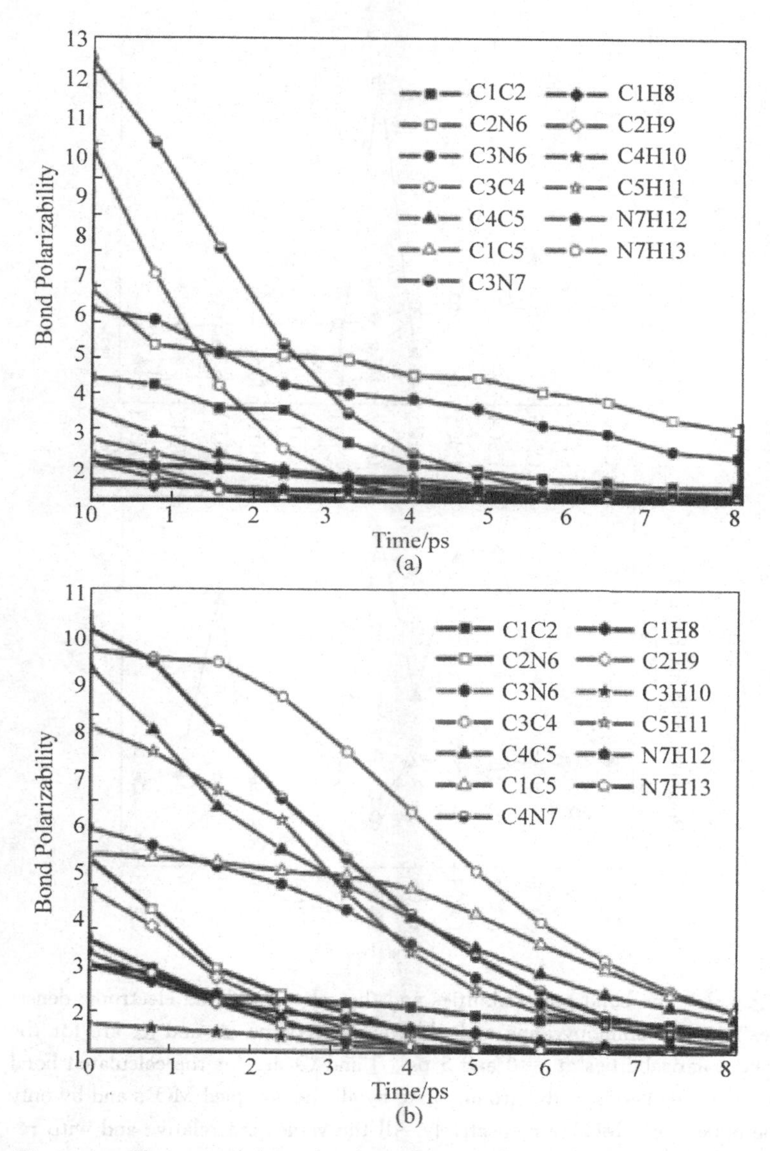

Fig. 4. 6 The temporal bond polarizabilities of (a) 2-aminopyridine and (b) 3-aminopyridine under 514. 5 nm excitation. The values for C3—C4 bond in (a) and C4—N7 bond in (b) at $t=0$ are normalized to 10, respectively.

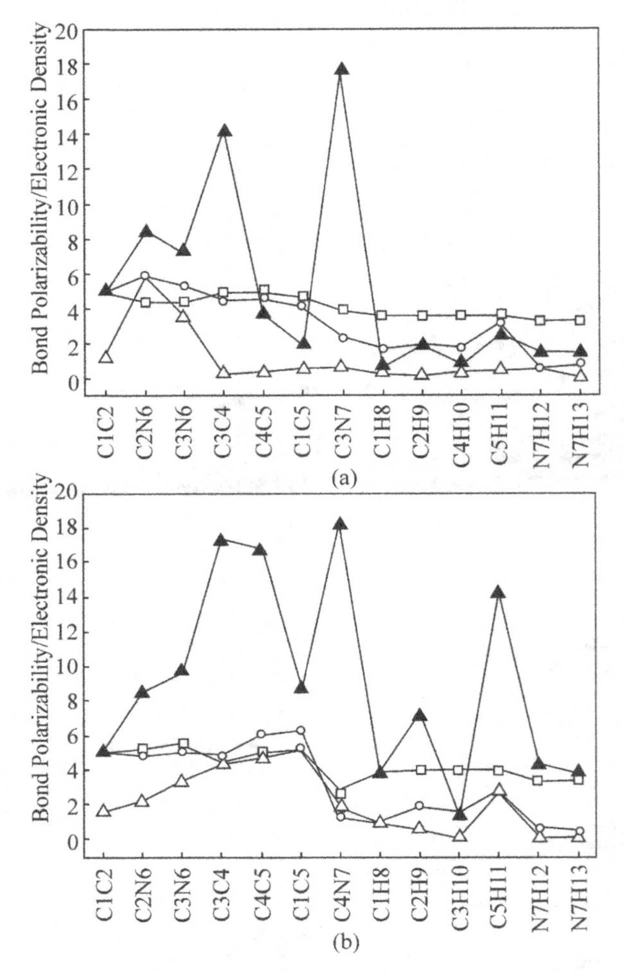

Fig. 4. 7 The bond polarizabilities and the calculated bond electronic densities for (a) 2-aminopyriding and (b) 3-aminopyridine. ▲ and △ are for the bond polarizabilities at $t=0$ and 5 ps. □ and ○ are for the calculated bond electronic densities of the ground state by all the occupied MO's and by only the upper eight MO's, respectively. All the values are relative and with respect to the value of C1—C2 bond which is normalized to 5 for convenience. Note that the profiles by △ are closer to those by ○ in both (a) and (b) cases. Note also that there is no comparability in values between the bond polarizabilities and the bond electronic densities.

We note that:

(1) The bond polarizabilities at the very initial stage of the birth of the virtual states are not consistent to the calculated bond electronic densities in any aspects. Their contrast is an indication of the electronic disturbance in the Raman process. The most impressive is that the bond polarizabilities of C3—N7 and C4—N7 in 2- and 3-aminopyridine respectively are very significant. This is depicted by the thicker sticks in Fig. 4. 5 along with other three bonds that also possess significant polarizabilities. On the contrary, the electronic densities of the ground state on these bonds are almost the least. (Note that they are the bonds where chemical breakage is mostly prone to occurring.) In particular, in 3-aminopyridine, the polarizabilities of the peripheral C5—H11 bond is significantly augmented. Also noted is that the bond polarizabilities of the other skeletal bonds are particularly small. The implication is that the excited charges tend to flow to the molecular periphery due to electronic repulsion. We believe that this is a generic property of Raman excited virtual states in many molecules.

(2) Though most relaxations of bond polarizabilities follow an exponentially decaying function, there are exceptions which can be followed by two decaying processes. Shown in Fig. 4. 8 are their samples. The decaying characteristic times, t_c, which were obtained through fitting by e^{-t/t_c} are shown in Fig. 4. 9. Though the relaxation behaviors of the bond polarizabilities are complicated, there is a notable point that in general the decaying characteristic times for the peripheral C—H and the branching C—N bond polarizabilities are shorter than those of the skeletal C—C or C—N bonds. This is particularly evident in the case of 3-aminopyridine. This is a logical consequence if the above assertion that the charges excited are flown toward the molecular periphery is recognized. Therefore, the relaxation which occurs in a reverse way is that the excitation in the peripheral C—H and the branching C—N bonds relaxes first with shorter decaying characteristic times. What may be puzzling is that those bond polarizabilities that relax with more than two processes are of the peripheral C—H/

N-H bonds in 2-aminopyridine, but are the skeletal C—C/C—N bonds in 3-aminopyridine. For these two-process relaxations, initially, the relaxation is slower with larger characteristic time and later, the relaxation is faster with smaller characteristic time.

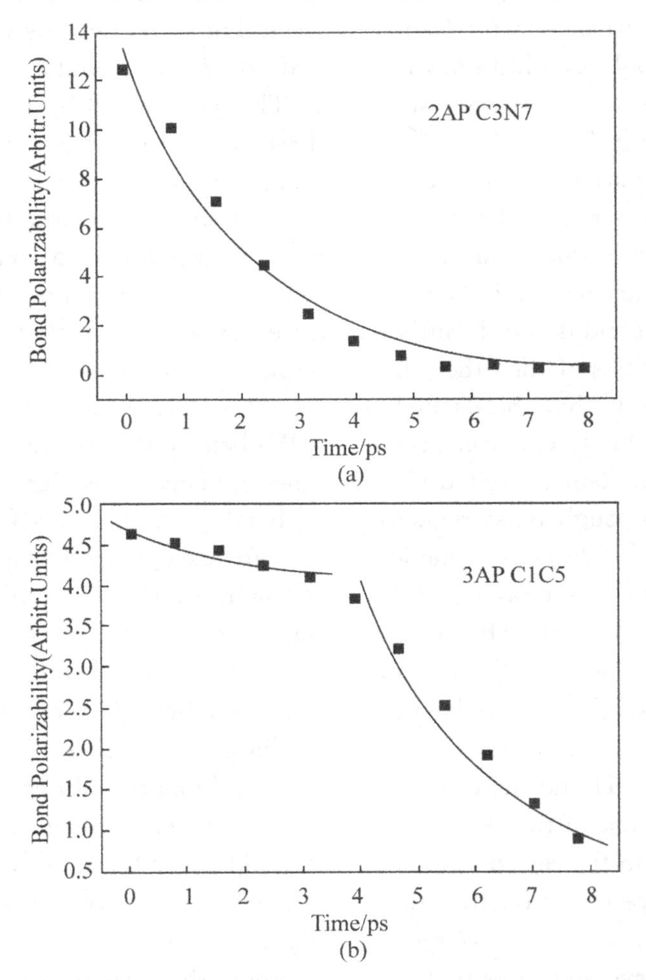

Fig. 4. 8 The relaxations with (a) one exponentially decaying process of the bond polarizability of C3—N7 of 2-aminopyridine and (b) two decaying processes for C1—C5 of 3-aminopyridine. The curves show the exponential function.

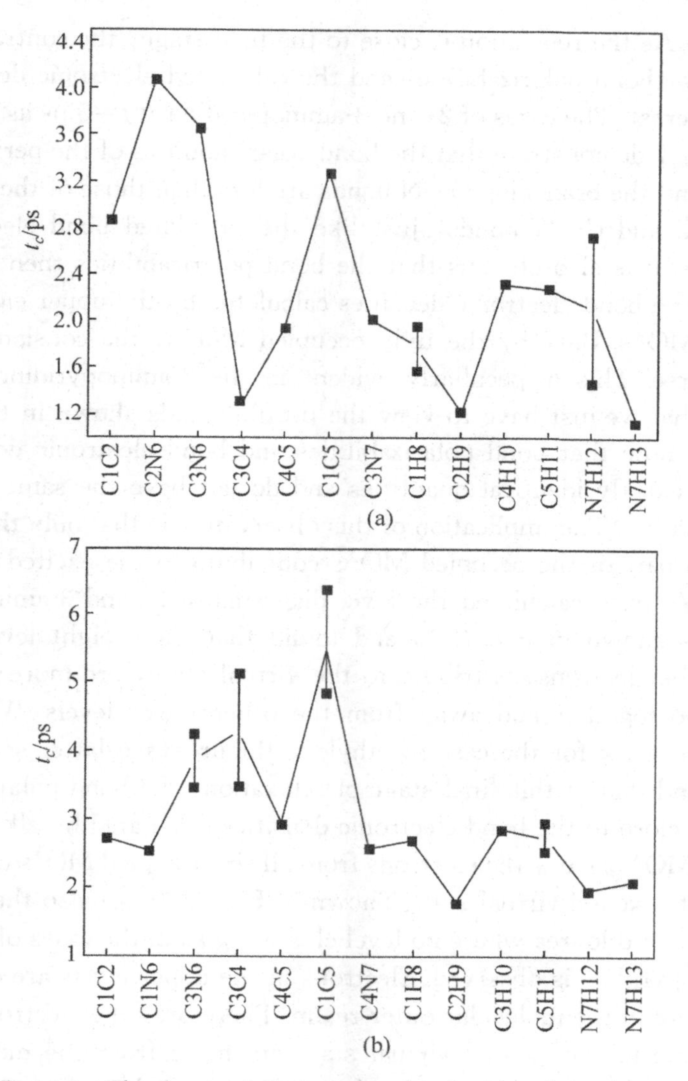

Fig. 4. 9 The characteristic times, t_c, for the relaxations of temporal bond polarizabilities of (a) 2-aminopyridine and (b) 3-aminopyridine. Note that some bond polarizabilities possess two decaying characteristic times.

(3) As the relaxation is close to the final stage, the contrast between the bond polarizabilities and the calculated electronic densities is of interest. The cases of 2- and 3-aminopyridine at $t=5$ ps as shown in Fig. 4. 7 demonstrate that the bond polarizabilities of the peripheral C—H and the branching C—N bonds are less than those of the skeletal C—C and C—N bonds, just like the calculated bond electronic densities. It is also obvious that the bond polarizabilities then do approach the bond electronic densities calculated by the upper eight occupied MO's. But, by the fully occupied MO's, the consistence is just worse. This is peculiarly evident in the 3-aminopyridine case. (Note that we just have to view the profile trends shown in the figure. We note that bond polarizabilities and bond electronic densities are not exactly identical quantities and do not have the same values numerically.)The implication of this observation is that only the electrons in part of the occupied MO's contribute to the excited virtual state. We have calculated the level diagrams of 2- and 3-aminopyridine, as shown in Fig. 4. 10 and found that these eight levels, of which the electrons contribute to the virtual state, are more or less clustered together and away from the other lower levels. We have done the work for the case of ethylene thiourea(see latter sections) and found that at this final stage of relaxation, the bond polarizabilities are close to the bond electronic densities calculated by all the occupied MO's, i. e. , the electrons from all the occupied MO's contribute to its excited virtual state. Shown in Fig. 4. 10 are also the levels of ethylene thiourea where no level clustering as in the cases of 2- and 3-aminopyridine is observed. Electrons in the upper levels are distributed more in the molecular outer realm. Therefore, the electrons that contribute to the excited virtual state are more from the molecular outer realm as demonstrated in the cases of 2- and 3-aminopyridine, though this is not the case for ethylene thiourea. This observation is important in our understanding of the charge excitation in the Raman virtual state.

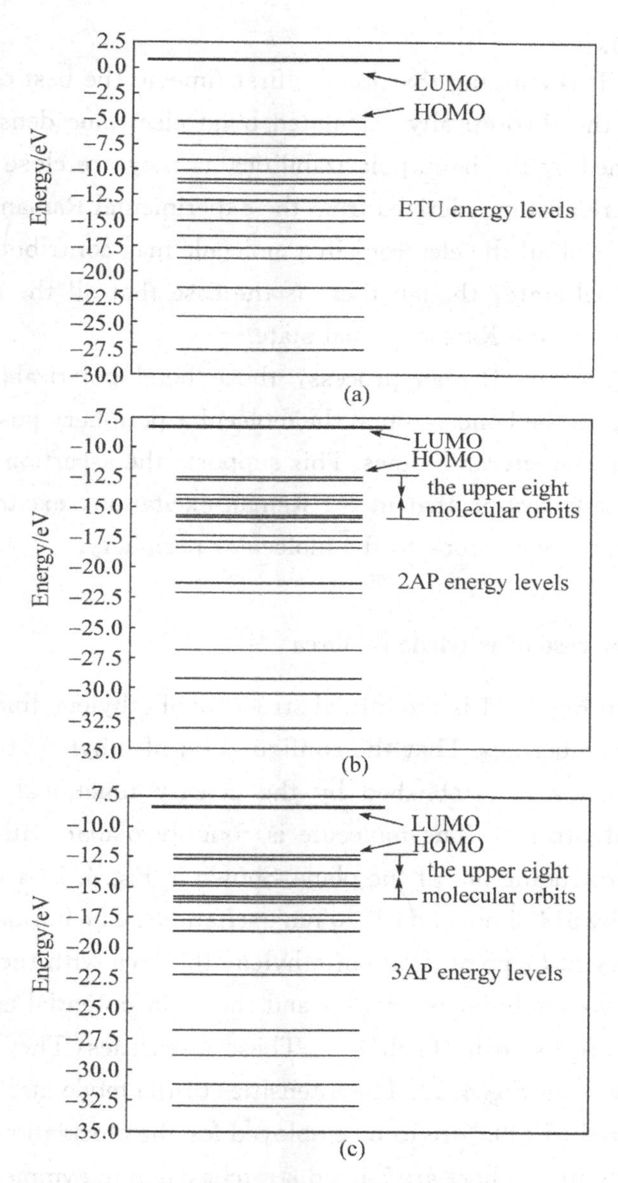

Fig. 4. 10 The molecular levels of (a) ethylene thiourea (ETU) (b) 2-aminopyridine (c) 3-aminopyridine.

Comments

(1) It is amazing that for the first time(to the best of our knowledge), the theoretically calculated bond electronic densities can be approached by the bond polarizabilities as they are close to the final stage of relaxation, derived from the experimental Raman intensities.

(2) Not all the electrons in a molecule may contribute to the Raman virtual state, though there is the case that all the electrons do contribute to the Raman virtual state.

(3) In the Raman process, those bond polarizabilities corresponding to the bonds around the molecular periphery possess smaller decaying characteristic times. This supports the assertion proposed in the previous section that in the Raman excitation, excited/disturbed charges are more prone to the molecular periphery.

4.3 The case of ethylene thiourea

Shown in Fig. 4. 11 is the formal structure of ethylene thiourea and its atomic numberings. That the configuration of ethylene thiourea is of C_2 symmetry is established by the density functional optimization (DFT algorithm). The molecule is roughly planar with two C—H bonds protruding out of the plane. Shown in Fig. 4. 12 are the Raman spectra by 514. 5 nm and 325. 0 nm excitations. The normal modes of a symmetry of C_2 point group of ethylene thiourea with their observed, fitted wavenumbers, intensities and the main potential energy distributions are listed in Table 4. 3. These intensities(They are also labeled by * in Fig. 4. 12. The intensities of the mode at 2965 cm^{-1} are normalized to 100.)are to be employed for the elucidation of the bond polarizabilities. There are 7 bond stretches(due to symmetry)for us to elucidate their bond polarizabilities. The requirement that bond stretch polarizabilities have to be positive greatly reduces the phase solution number to 6. We are very fortunate that even among these multiple solutions, the corresponding bond stretch polarizabilities are very similar. Thus, the most crucial step by this algorithm is solved.

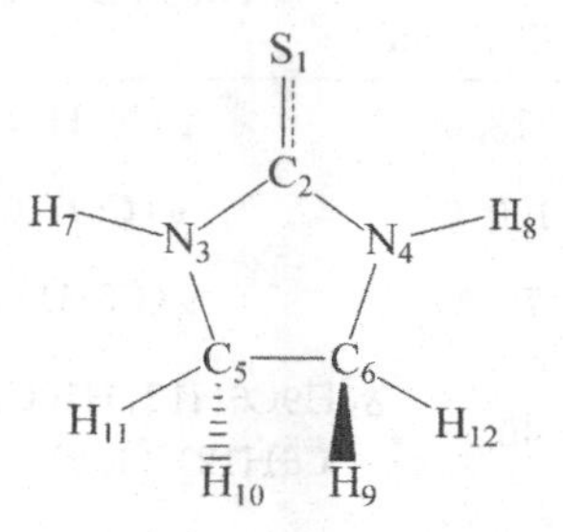

Fig. 4. 11 The formal structure of ethylene thiourea and its atomic numberings. It is roughly planar with C5—H10 and C6—H9 bonds more protruding out of the molecular plane.

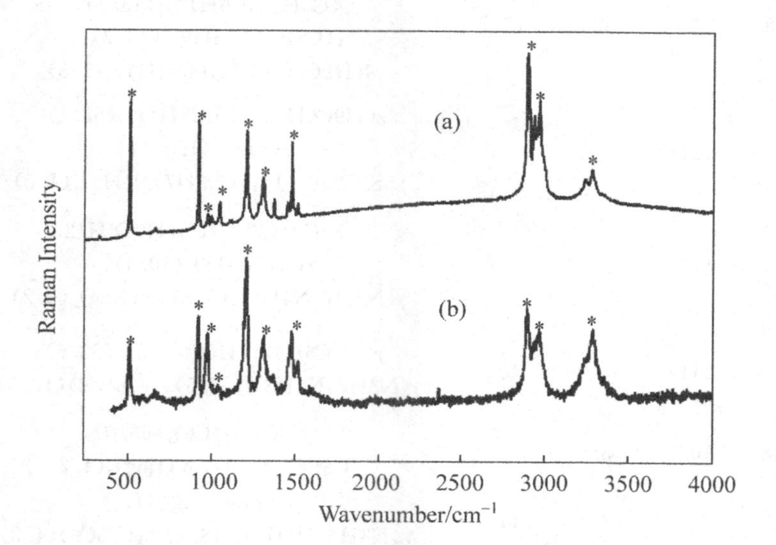

Fig. 4. 12 The Raman spectra of ethylene thiourea by (a) 514. 5 nm and (b) 325 nm excitations. * shows the peaks that appear in Table 4. 3 and are employed for the elucidation of the bond polarizabilities.

Table 4.3 The normal modes of A symmetry of C_2 point group of ethylene thiourea with their observed, fitted wavenumbers, intensities and main potential energy distributions. The intensities of the mode at 2965 cm^{-1} are normalized to 100. ν stands for bond stretching. δ, γ and τ stand for the bending-type coordinates.

Raman Shift/cm^{-1}		Intensity		Potential Energy Distribution/%
Exp.	Fitted	514.5nm	325.0nm	
3286	3279	30.0	113.6	ν(N3H7,N4H8)$_s$(100.0)
2965	2959	100.0	100.0	ν(C5H10,C6H9)$_s$(93.9)
2896	2886	88.0	75.3	ν(C5H11,C6H12)$_s$(92.6)
1490	1498	30.1	40.4	δ(H9C6H12,H10C5H11)$_s$(77.2), ν(C5H11, C6H12)$_s$(1.9), ν(C5H10,C6H9)$_s$(1.9)
1309	1311	30.3	46.4	δ(N3H7,N4H8)$_s$(77.9), ν(C2N3,C2N4)$_s$(1.7), ν(N3C5,N4C6)$_s$(1.4), ν(C5C6)(1.2)
1208	1211	20.5	70.7	ν(C2N3,C2N4)$_s$(37.6), ν(C5H11,C6H12)$_s$(13.7), ν(C5H10,C6H9)$_s$(13.7), δ(H9C6H12,H10C5H11)$_s$(1.3)
—	1193	0.0	0.0	τ(H9C6H12,H10C5H11)$_s$(52.1), τ(ring)$_s$(12.4), γ(S1C2)(1.1), γ(N3H7,N4H8)$_s$(1.0)
—	1154	0.0	0.0	γ(S1C2)(37.6), γ(H9C6H12, H10C5H11)$_s$(10.4), γ(N3H7,N4H8)$_s$(9.8), τ(ring)$_s$(2.2)
—	1112	0.0	0.0	γ(H9C6H12,H10C5H11)$_s$(33.6), γ(N3H7,N4H8)$_s$(14.5), γ(S1C2)(10.3)
1048	1058	3.5	0.2	ν(N3C5,N4C6)$_s$(68.0), ν(C5C6)(16.9), δ(ring)$_s$(1.2)
977	967	3.1	44.4	δ(ring)$_s$(52.4), δ(N3H7,N4H8)$_s$(13.8), ν(C5C6)(1.3)
922	928	19.4	36.5	ν(C5C6)(57.4), ν(N3C5,N4C6)$_s$(10.0), δ(ring)$_s$(1.8)

—	574	0.0	0.0	γ(N3H7,N4H8)$_s$(64.4), γ(S1C2)(4.1), γ(H9C6H12,H10C5H11)$_s$(2.2), τ(ring)$_s$(1.0),
508	505	33.5	34.6	ν(S1C2)(69.9), ν(C2N3, C2N4)$_s$(1.5), δ(ring)$_s$(1.1)
—	225	0.0	0.0	τ(ring)$_s$(72.2), τ(H9C6H12, H10C5H11)$_s$(1.1)

Shown in Fig. 4. 13 are the bond polarizabilities of ethylene thiourea at the initial excitation moment and the final stage of relaxation for both 514. 5 nm and 325 nm excitations, together with the calculated bond electronic densities of the ground state. Shown in Fig. 4. 14 are the bond polarizabilities under 514. 5 nm and 325 nm excitations as a function of time. It shows the relaxation of the Raman excited virtual states. From these, we have the following observations.

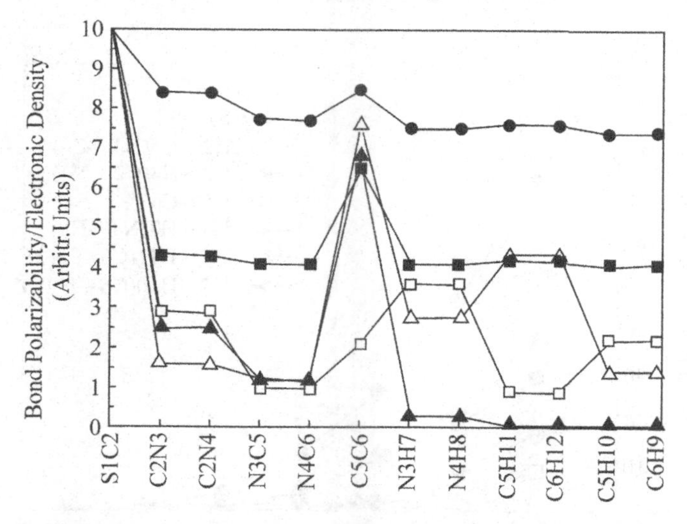

Fig. 4. 13 The bond polarizabilities of ethylene thiourea at the initial excitation moment(△ and □ for 514. 5 nm and 325 nm excitations, respectively) and the final stage of Raman relaxation(▲ and ■ for 514. 5 nm and 325 nm excitations, respectively.). ● is for the calculated bond electronic densities of the ground state. The value of S1C2 bond is normalized to 10.

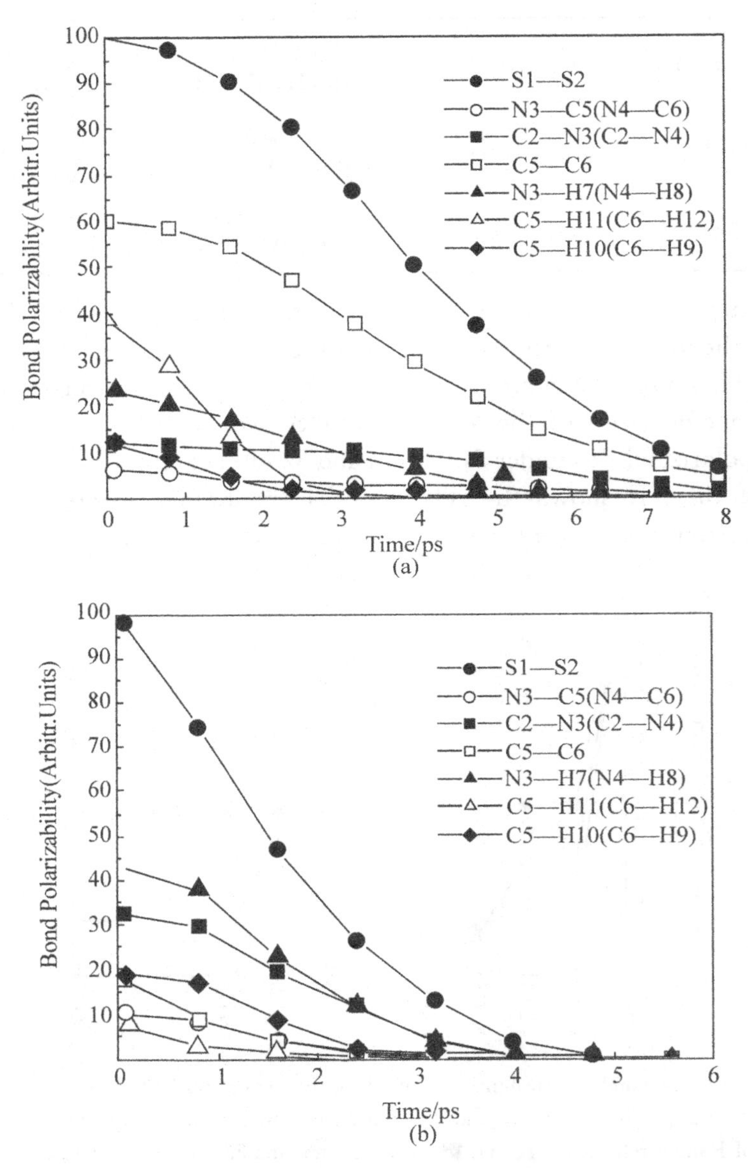

Fig. 4. 14 The bond polarizabilities of ethylene thiourea under (a) 514. 5 nm and (b) 325 nm excitations as a function of time. The value for S1C2 bond at $t=0$ is normalized to 100.

(1) At the initial excitation moment by the 514. 5 nm excitation, the tendency of the bond polarizabilities is closer to the bond electronic densities of the ground state than that by the 325 nm excitation. This implies the less electronic disturbance by the 514. 5 nm excitation. By the 325 nm excitation, the situation changes much. Firstly, the bond polarizabilities of C2—N3 (and C2—N4) and N3—H7 (and N4—H8 which are in the peripheral) are enhanced greatly. The flow of electrons from S1—C2 to these bonds is significant. Secondly, the bond polarizability of C5—C6 decreases. Meanwhile, the polarizabilities of two C—H bonds (C5—H10 and C6—H9), which protrude out of the molecular plane and are hence less crowded to each other, are greater than those of the roughly in-plane C5—H11 and C6—H12 bonds. This situation is opposite in the 514. 5 nm excitation. Our interpretation is that under the Raman excitation, the excited electrons tend to flow to the molecular peripheral, especially those bonds that are less crowded, due to the electronic repulsion and this effect is more obvious under the more energetic 325 nm excitation.

Under both 514. 5 nm and 325 nm excitations, most peripheral N—H and C—H bonds possess larger polarizabilities than the skeletal C2—N3 (C2—N4) and N3—C5 (N4—C6) bonds. This conclusion that the excited electrons tend to flow to the molecular peripheral due to electronic repulsion is believed to be a generic property of the Raman excited states.

(2) As time evolves, for the 514. 5 nm excitation, the bond polarizability of C5—H11 (C6—H12) declines much faster than those of the remaining bonds (Fig. 4. 14). This is notable. Furthermore, the decay rates for the bond polarizabilities of N3—H7 (N4—H8) and C5—H10 (C6—H9) in the molecular peripheral are similar. So are the bond polarizabilities of the adjacent C2—N3 (C2—N4) and N3—C5 (N4—C6) bonds. For the 325 nm excitation, the decay behaviors of the bond polarizabilities of C5—C6 and C5—H11 (C6—H12), in the molecular plane, are similar and those of the rest bonds (except S1—C2) are similar to each other at least in the first 5 ps.

(3) When the final stage of relaxation is approached, as shown in Fig. 4. 13, the relative magnitudes of the bond polarizabilities are approaching the bond electronic densities of the ground state by the quantum chemical calculation despite of 514. 5 nm and 325 nm excitations. We note that for the 514. 5 nm excitation, in the final stage of relaxation, the bond polarizabilities of the peripheral N—H and C—H bonds are less than those of the skeletal C2—N3(C2—N4) and N3—C5(N4—C6) bonds, close to the result by the quantum chemical calculation for the ground state. This is in contrast with the situation at the very initial moment of the virtual state that the peripheral N—H and C—H bonds possess larger polarizabilities. This demonstrates vividly the relaxation of the excited electrons from the peripheral N—H and C—H bonds toward the inner skeletal C—N bonds. The relaxation recovers the electronic distribution toward that of the ground state. This is also true under the 325 nm excitation. That we can observe the (relative) bond electronic densities of the molecular ground state via the Raman spectra, which previously were merely theoretical quantities by the quantum chemical calculation based on the idea of Schroedinger's wavefunction as noted in the last section [see Comment (1) of Section 4. 2] is confirmed here again.

(4) Finally, it is noted that the relaxation of the temporal bond polarizabilities are in the range around 8 ps and 5 ps, respectively, by 514. 5 nm and 325 nm excitations as shown in Fig. 4. 14. These time durations show roughly the lifetimes of Raman virtual states under 514. 5 nm and 325 nm excitations. This ratio of 8 : 5 is roughly proportional to the wavelength ratio of the excitations, 5 : 3. This is in agreement with Heisenberg's uncertainty principle that lifetime is inversely proportional to the excitation energy. That is, lifetime is proportional to the wavelength of the exciting light. This confirms that Raman excited states are not stationary eigenstates. Hence, it is proper to call these states non-stationary or *virtual*.

Comments

The case of ethylene thiourea shares common results with those of 2-amino and 3-aminopyridine. These demonstrate the common properties of Raman excited virtual states.

4.4 The case of ethylene thiourea adsorbed on the silver electrode

Surface enhanced Raman scattering (SERS), which is a phenomenon that adsorbed molecules on the roughened metal (like Ag, Au) surfaces can have an enhanced Raman cross section up to a million fold, has been applied to surface and interfacial studies, including very sensitive single molecule detection[1-3]. Currently, there are two proposed mechanisms for SERS[4-6]. One is called the charge transfer mechanism due to the charge transfer between the adsorbed molecule and the metal surface, and the other the electromagnetic mechanism due to the electromagnetic field very locally enhanced around the site where the adsorbed molecule resides One of the prominent phenomena in SERS is its intensities are surface potential dependent. For the retrieval of the physical background behind this unique behavior, we will try to elucidate the bond polarizabilities from the SERS intensities. The compound that we will try on is ethylene thiourea adsorbed on the Ag electrode.

The structure ofethylene thiourea was shown in the last section. For convenience, it is reshown in Fig. 4. 15, including its atomic numberings. As mentioned before, the molecule is roughly planar with two C—H bonds protruding out of the plane. It possesses C_2 symmetry. That it is bound to the Ag surface through its S atom with a vertical configuration is implied by its SERS spectrum. In fact, the proposition of the adsorption configuration is not so crucial in our algorithm. The bonding of the adsorbed molecules to the surface is weak and is, in general, less than one tenth that of a covalent bond[7, 8]. Hence, we may just treat the adsorbed ethylene thiourea itself as an entity without surface atoms and do its normal mode analysis. This is justifiable by the observation that the Ag-S mode frequency (210 cm^{-1}) is much less than those intramolecular vibrational modes, showing the weak coupling between the adsorbed ethylene thiourea molecule and the Ag surface, and that this mode intensity is also much less than those due to the intramolecular motion.

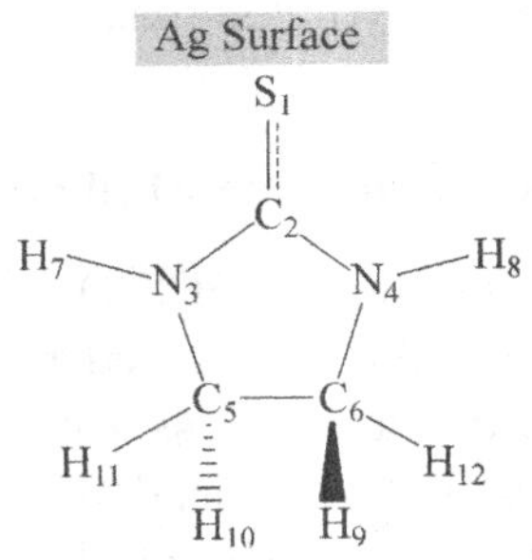

Fig. 4. 15 The formal structure of ethylene thiourea and its atomic numberings. It is roughly planar with C5—H10 and C6—H9 bonds more protruding out of the molecular plane. The adsorption on Ag surface is through S atom with a vertical configuration.

We know that bond polarizabilities are derived from SERS intensities, hence, they will encompass all the adsorption effects through the SERS process. Due to the possible heterogeneity of the surface structure, the bond polarizabilities obtained are the average over the surface area studied.

Shown in Fig. 4. 16 are the SERS spectra of ethylene thiourea under various potentials by 514. 5 nm excitation. Under various potentials, SERS peak positions may shift somewhat. However, it is found that even when this deviation is up to 20 cm^{-1}, the L_{tj} elements remain quite unchanged. Thus, we can have one $[L_{tj}]$ matrix for all the treatments under different potentials.

We will be mainly concerned with the polarizabilities of bond stretches. This is because that these bond polarizabilities are significantly larger than those of the bending coordinates and will show more physical significance. Therefore, for the elucidation of the polarizabilities of the 7 bond stretches (The number of independent bond stretches is reduced to 7 because of C_2 symmetry.), we need 7 mode intensities whose modes are mostly consisted of bond stretches. These SERS peaks are labeled by * as shown in Fig. 4. 16. These mode wavenumbers (at −0. 9 V, all voltages are with respect to the stand-

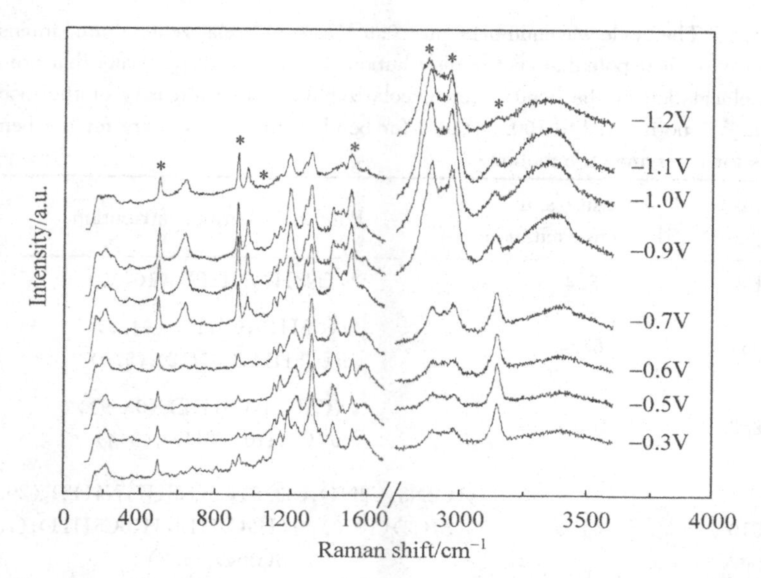

Fig. 4. 16 The SERS spectra of ethylene thiourea by 514. 5 nm excitation under various potentials(voltage is with respect to the standard calomel electrode). Those 7 peak intensities that are employed for the elucidation of bond(stretch)polarizabilities are labeled by *.

ard calomel electrode, abbreviated as SCE), (integrated) intensities together with their potential energy distributions(PED)are shown in Table 4. 4. (At this point, one may be puzzled by the seemingly significant components of bending motion in the mode at 1510 cm^{-1}, for instance. However, the error caused may not be necessarily significant since we are employing altogether 7 mode intensities to elucidate the bond polarizabilities. In fact, we note that our algorithm relies on a set of mode intensities, instead of an individual one, for the bond polarizability analysis and is therefore robust against intensity uncertainty in one or two mode intensities, say.)We are lucky that the requirement that the polarizabilities of the 7 bond stretches have to be of the same sign leads to a unique phase set. Thus, the phase problem is solved. Shown in Fig. 4. 17 are these bond polarizabilities under various potentials at the very initial moment of Raman excitation.

Table 4.4 The mode wavenumbers (at −0.9 V, SCE), relative integrated intensities together with their potential energy distributions of the 7 spectral peaks that are used for the elucidation of the bond (stretch) polarizabilities. The intensity of the mode at 2883 cm^{-1} is normalized to 100. ν stands for bond stretch. τ, δ, γ are for the bending and s is for the symmetric motion.

Raman Shift/cm^{-1}	Intensity/a.u. 514.5nm excitation	Potential Energy Distribution
3139	5.4	ν(N3H7, N4H8)$_s$ (100.0)
2969	68.4	ν(C5H10, C6H9)$_s$ (94.9), ν(C5H11, C6H12)$_s$ (5.0)
2883	100.0	ν(C5H11, C6H12)$_s$ (93.9), ν(C5H10, C6H9)$_s$ (5.4)
1510	41.3	ν(C2N3, C2N4)$_s$ (33.6), δ(N3H7, N4H8)$_s$ (20.6), ν(S1C2) (10.6), τ(H9C6H12, H10C5H11)$_s$ (7.5), δ(ring)$_s$ (5.2)
1037	0.1	ν(N3C5, N4C6)$_s$ (73.0), ν(C5C6) (11.7), δ(ring)$_s$ (4.7)
917	5.5	ν(C5C6) (57.1), ν(N3C5, N4C6)$_s$ (6.1), δ(ring)$_s$ (10.57)
495	4.6	ν(S1C2) (49.8), γ(N3H7, N4H8) (26.4)

From Fig. 4.17, the behavior of the bond polarizabilities can be categorized into two regions: in between −0.3 V and −0.7 V, where all the bond polarizabilities are small and do not vary significantly on the applied voltage, and beyond −0.7 V to −1.2 V, where all the bond polarizabilities are enhanced significantly. In the former region, we suppose that the electromagnetic mechanism plays the major role since it shows the distance effect: the closer a bond is to the adsorption site, the larger is its bond polarizability[9]. This is indicated by that the bond polarizability of N3—H7 (N4—H8) is the largest. Also, the bond polarizability of C2—N3 (C2—N4) is larger than that of N3—C5 (N4—C6). (That the bond polarizability of S1—C2 is not so large can be attributed to the binding effect on the metal sur-

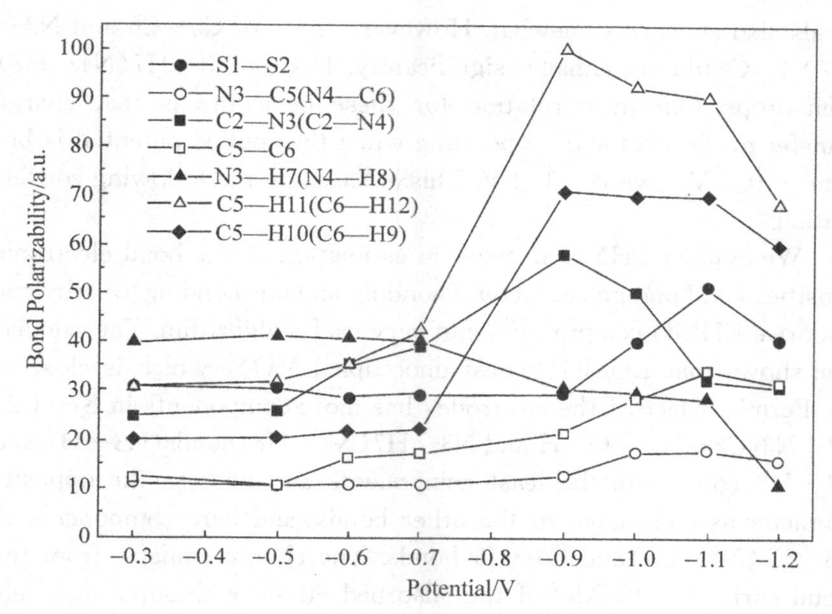

Fig. 4. 17 The bond polarizabilities of ethylene thiourea under various potentials (SCE) at the very initial moment of SERS excitation. The bond polarizability of C5—H11, at −0. 9 V is normalized to 100.

face.) Furthermore, the electromagnetic effect also implies that a bond that is normal to the electric field (hence, parallel to the surface) will have a polarizability smaller than that of the other bond (of the same type) that is parallel to the field (hence normal to the surface) (Ref[10], Chapter 5 Section 5. 3). If the vertical adsorption configuration through the S—Ag bonding is adopted, the C5—H11(C6—H12) bond is parallel to the electric field (normal to the surface) and the C5—H10(C6—H9) bond is normal to the electric field (parallel to the surface), more or less. This inference leads exactly to what we have that the bond polarizability of C5—H10(C6—H9) is smaller than that of C5—H11(C6—H12).

As the potential is more negative to −0. 7 V, the polarizabilities of C—H bonds (including both C5—H10 (C6—H9) and C5—H11 (C6—H12)) enhance greatly. Those of C2—N3(C2—N4) and S1—C2 bonds also enhance somewhat. However, those of C5—C6 and N3—C5(N4—C6) do not enhance significantly. That of N3—H7(N4—H8) even drops. Our interpretation for these behaviors is that charge transfer mechanism starts operating when the applied potential is beyond −0. 7 V toward −1. 2 V. This is based on the following consideration.

We applied EHMO to have an estimation of the bond electronic densities and bonding characters(bonding and antibonding) of ethylene thiourea. EHMO is a primitive but very useful algorithm. The calculation shows that LUMO(lowest unoccupied MO), which is close to the Fermi surface of the electrode, has more components in S1—C2, C2—N3(C2—N4), C—H and N3—H7(N4—H8) bonds(N3—H7 and N4—H8 contribute the least components and possess the opposite character as contrasted to the other bonds) and bare components in N3—C5(N4—C6) and C5—C6 bonds. The charge transfer from the metal surface to LUMO of the adsorbed ethylene thiourea molecule for the SERS mechanism can explain the above bond polarizability behaviors(as shown in Fig. 4. 17). The bond polarizabilities of S1—C2, C2—N3(C2—N4), C—H bonds are enhanced since they have more charges to involve in this mechanism. While there are none/few charges in N3—C5(N4—C6) and C5—C6 bonds to operate to enhance their polarizabilities. This consideration may also explain the suppression of the bond polarizability of N3—H7(N4—H8) when the potential is more negative to −0. 7 V, due to its bare component in LUMO and opposite bond character. (We note that the fourth MO below HOMO possesses similar characters as LUMO. Its involvement in SERS mechanism by charge transfer to the electrode is possible. Since LUMO is close to the Fermi surface, LUMO may play the major role.)

It deserves attention that, as shown in Fig. 4. 17, the difference between the bond polarizabilities of C5—H11(C6—H12) and C5—H10(C6—H9) across the potential range from −0. 3 V to −1. 2 V remains quite the same. This implies that the electromagnetic effect may persist toward −1. 2 V even when the charge transfer effect starts operating.

For the temporal bond polarizabilities, we found that in all cases, there is but *one* exponentially decaying behavior despite of the various bonds and under the various applied potentials. This is impressive and the case for C5—H11 is shown in Fig. 4. 18 for demonstration. The fitted characteristic times, t_c, by e^{-t/t_c} function for the relaxations of the various temporal bond polarizabilities under various potentials are shown in Fig. 4. 19, from which we have the following observations:

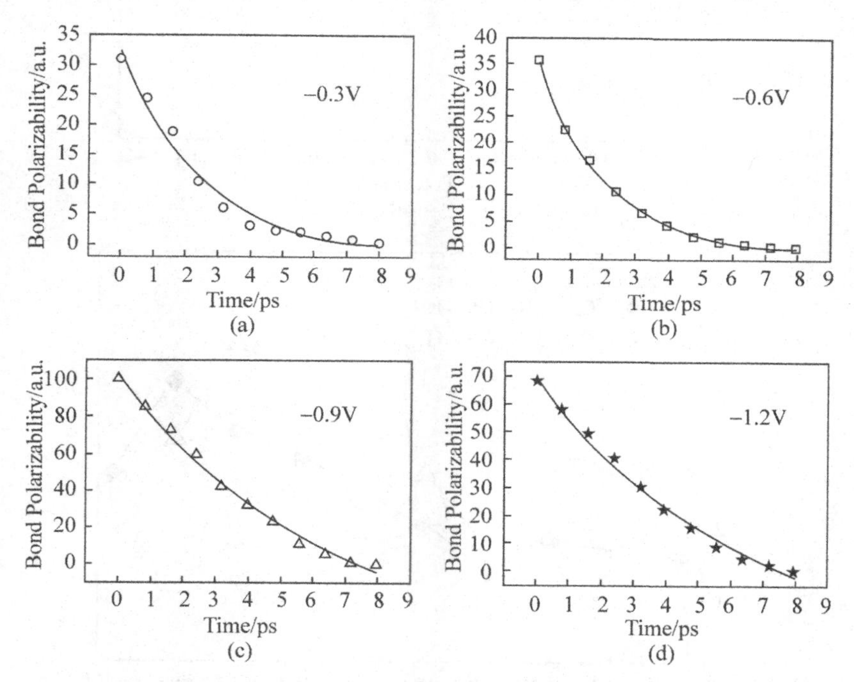

Fig. 4. 18 The fitted exponentially decaying curves for the temporal bond polarizabilities(discrete data)of C5—H11 at various potentials(SCE).

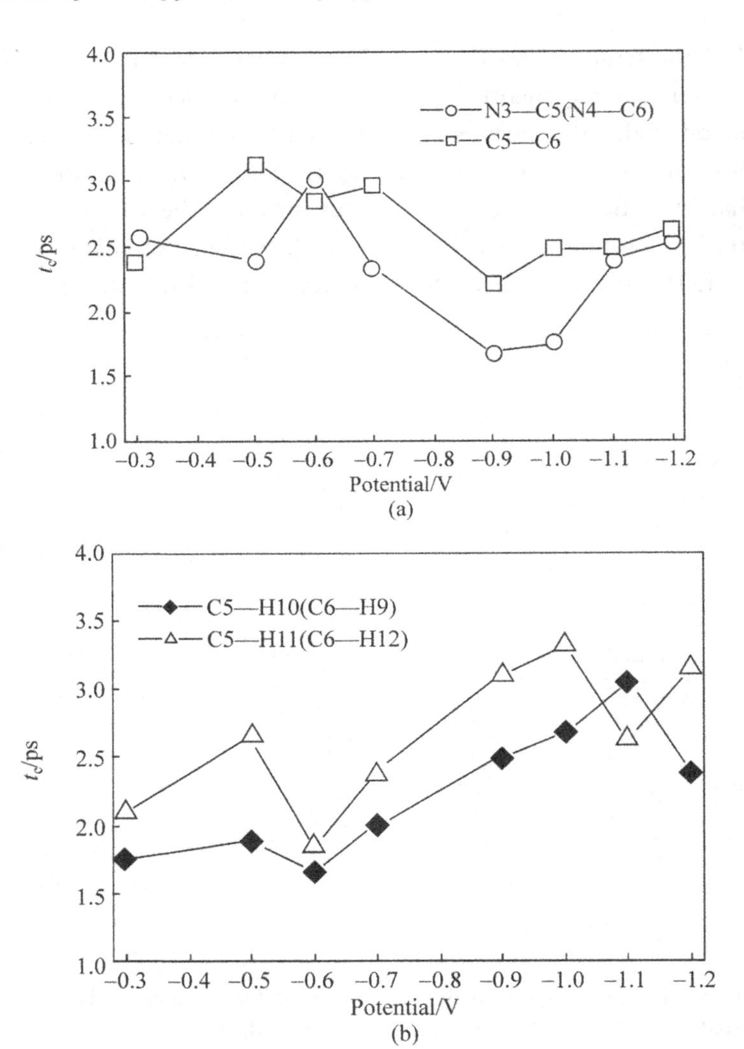
4.0
3.5
3.0
2.5
2.0
1.5
1.0
t_c/ps
—o— N3—C5(N4—C6)
—□— C5—C6
−0.3 −0.4 −0.5 −0.6 −0.7 −0.8 −0.9 −1.0 −1.1 −1.2
Potential/V
(a)
—◆— C5—H10(C6—H9)
—△— C5—H11(C6—H12)
Potential/V
(b)

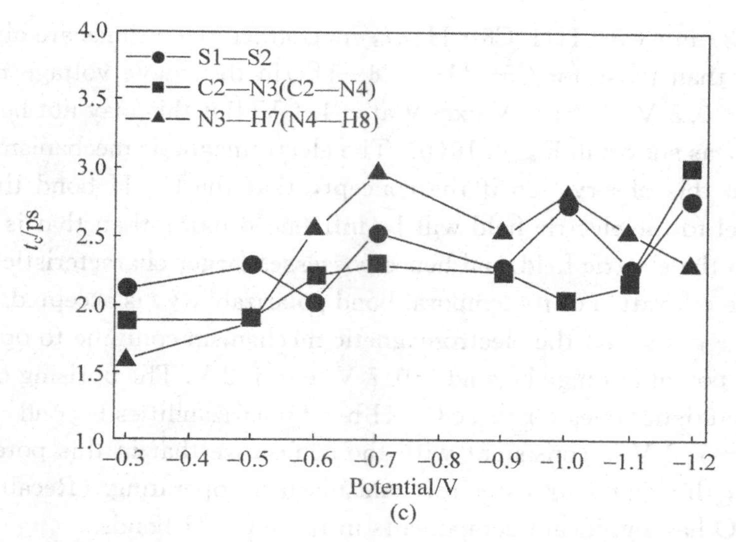

Fig. 4. 19 The characteristic times, t_c, for the relaxations of the various temporal bond polarizabilities of ethylene thiourea at various potentials(SCE).

(1) For N3—C5(N4—C6) and C5—C6 bonds, as shown in Fig. 4. 19(a), we do observe that their characteristic times drop somewhat as the applied potential swifts across −0. 7 V toward −1. 2 V, in which the charge transfer mechanism starts operating. This is consistent with the previous argument that these bonds are least involved in the charge transfer mechanism, due to their least charge densities in LUMO. The reason is that the characteristic time of the relaxation due to the charge transfer mechanism is certainly larger(than that via the electromagnetic mechanism) since this relaxation needs longer time for the charges to re-distribute. Hence, the less a bond is involved in the charge transfer mechanism, the shorter the characteristic time of its temporal bond polarizability will be. This will be more evident as compared with the other bond polarizabilities which possess larger characteristic times as the applied potential swifts across −0. 7 V toward −1. 2 V as shown below.

(2) For C5—H11(C6—H12), its characteristic times are always larger than those for C5—H10(C6—H9) in the whole voltage range from −0.3 V to −1.2 V except at −1.1 V(But this may not be crucial.), as shown in Fig. 4.19(b). The electromagnetic mechanism will lead to this observation if the concept, that the C—H bond that is parallel to the electric field will be influenced more(than that is normal to the electric field) and hence possesses larger characteristic time for the relaxation of its temporal bond polarizability, is accepted. This also suggests that the electromagnetic mechanism continue to operate in the potential range beyond −0.7 V to −1.2 V. The uprising of the characteristic times for these C—H bond polarizabilities beyond −0.7 V to −1.2 V is consistent with the conjecture that in this potential range, the charge transfer mechanism starts operating. (Recall that LUMO has significant components in these C—H bonds.)

(3) For S1—C2, C2—N3(C2—N4) and N3—H7(N4—H8), as shown in Fig. 4.19(c), the characteristic times of their temporal bond polarizabilities increase in general from −0.3V to −1.2V. This is consistent with the conjecture that SERS shifts from the electromagnetic mechanism to, mainly, the charge transfer mechanism as the potential is toward −1.2V. The characteristic times for the relaxations of these bond polarizabilities increase accordingly, since these bonds have more components in LUMO to involve in the charge transfer mechanism. As mentioned above, the charge transfer mechanism will lengthen the characteristic time of the relaxation of a temporal bond polarizability. It is also interesting to note that the characteristic time for S1—C2 does not vary so significantly. This may be due to its strong binding to the metal surface in the whole potential range.

(4) The bond polarizabilities corresponding to complete relaxation are shown in Fig. 4.20. Those of ethylene thiourea in the crystalline form are also attached. They are consistent with the bond densities in the ground state as has been shown in Fig. 4.13. However, the contrast shows that the bond polarizability of C2—N3(C2—N4) is significantly augmented in this SERS case. (Very probably, that of S1—

C2 is also enhanced. Note that only the relative bond polarizabilities are obtained in this work.) Apparently, this can be attributed to the adsorption effect which results in the aggregation of charges. That only the polarizability of the C2—N3(C2—N4) bond is affected significantly demonstrates that the adsorption effect is of very short distance. (If one is careful enough, he may sense that the bond polarizability of N3—C5(N4—C6) is more depressed in the SERS case than in the crystalline form due to that its charge is shifted toward C2—N3 (C2—N4) because of adsorption.) We have to appreciate this result since it offers us a very first hand information of the adsorption effect. This would be very hard to obtain by other experimental methods or even theoretical calculation.

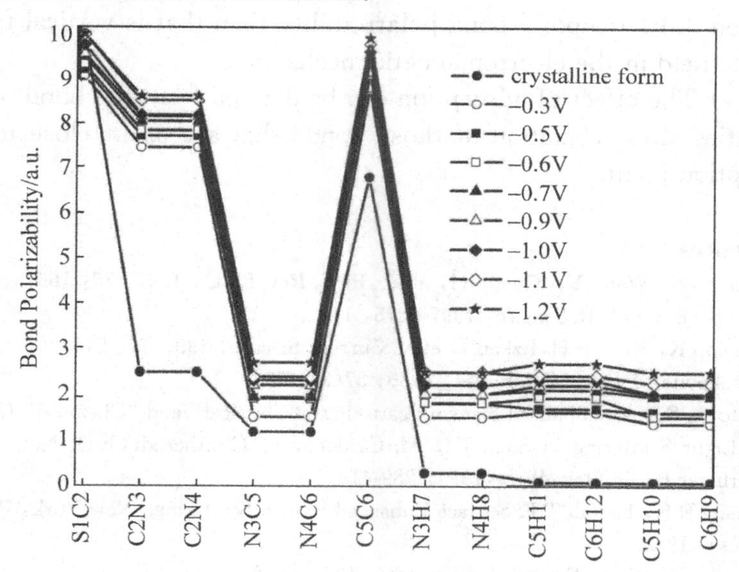

Fig. 4. 20 The bond polarizabilities of ethylene thiourea after complete relaxation under various potentials(SCE) in SERS and those in the crystalline form. Those of the S1—C2 bond at —1. 2 V and in the crystalline form are normalized to 10.

Comments

We have the following conclusions for this study:

(1) From the bond polarizabilities, as derived from Raman intensities, we may infer that there are two mechanisms involved in SERS.

(2) The characteristic time of the relaxation of the bond polarizability due to charge transfer mechanism is certainly larger than that via the electromagnetic mechanism since this relaxation needs longer time for the charges to re-distribute. Hence, the less a bond is involved in the charge transfer mechanism, the shorter the characteristic time of its temporal bond polarizability will be.

(3) The bond that is parallel to the electric field will be influenced more, and hence possesses larger characteristic time for the relaxation of its temporal bond polarizability than that is vertical to the electric field in the electromagnetic mechanism.

(4) The effect of adsorption can be derived from the bond polarizabilities after relaxation of those bonds that are on or close to the adsorption point.

References

[1] Kneipp K, Wang Y, Kneipp H, et al. Phys. Rev. Lett., 1997, 78: 1667.
[2] Nie S, Emory S R. Science, 1997, 275: 1102.
[3] Kneipp K, Kneipp H, Itzkan I, et al. Current Science, 1999, 77: 915.
[4] Moskovits M. Rev. Mod. Phys., 1985, 57(3): 783.
[5] Otto A. Surface-enhanced Raman Scattering: "Classical" and "Chemical" Origin, in Light Scattering in Solid IV, M. Cardona, G. Guntherodt (Ed). New York: Springer Berlin Heidelberg, 1984: 289-418.
[6] Chang R K, Furtak T E. Surface Enhanced Raman Scattering. New York: Plenum Press, 1982.
[7] Huang Y, Wu G. Spectrochimica Acta, 1989, 45A: 123.
[8] Wang P, Wu G. Chem. Phys. Letters, 2004, 385: 96.
[9] Qi J, Wu G. Spectrochimica Acta, 1989, 45A: 711.
[10] Tian B, Wu G, Liu G. J. Chem. Phys., 1987, 87: 7300.

Chapter 5
More applications

5.1 The case of methylviologen and its adsorption on the silver electrode

Methylviologen(MV) possesses D_{2h} symmetry with a nearly planar structure. It has 15 modes of A_g symmetry. Fig. 5.1 shows it structure and atomic numberings. Fig. 5.2 is its Raman spectrum(solid $MVCl_2$) under 514.5 nm excitation. We will treat the three C—H bonds of the methyl group equivalently. This leads to eight independent bond stretching polarizabilities to figure out. Among the 15 modes of A_g symmetry, there are eight modes that are mostly consisted of the stretching coordinates. The intensities of these eight modes will be employed for the analysis. They are labeled by * in Fig. 5.2. In the normal mode analysis, the symmetry coordinates were employed. The eight symmetry coordinates of which the bond polarizabilities are to be elucidated are tabulated in Table 5.1 in terms of the bond stretching coordinates. Once the polarizabilities corresponding to these symmetry coordinates are deduced, the bond polarizabilities can be obtained readily by symmetry. Normal mode analysis was tried by refining the force constants which were initially simulated by DFT with ub3lyp/cc-pvDZ till the calculated mode frequencies and the measured ones were consistent. The corresponding observed wavenumbers, fitted wavenumbers, relative intensities(integrated over wavenumber) and PED are tabulated in Table 5.2.

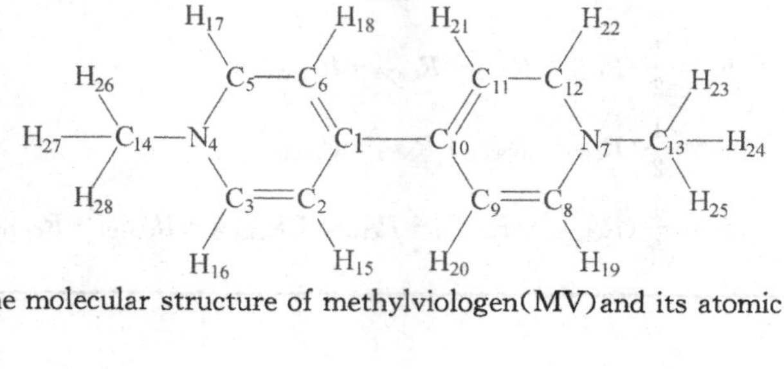

Fig. 5.1 The molecular structure of methylviologen(MV) and its atomic numberings.

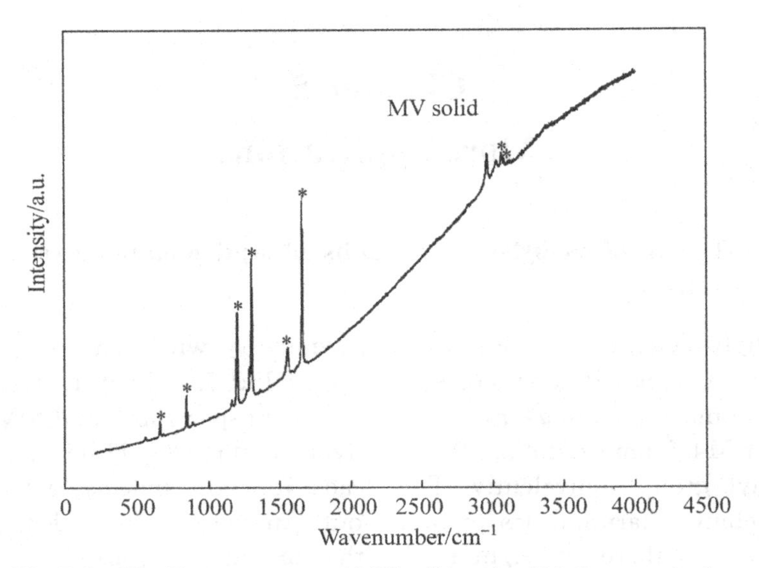

Fig. 5. 2 The Raman spectra of MV solid under 514. 5 nm excitation. The eight peak intensities employed for the elucidation of bond(stretch) polarizabilities are labeled by * .

Table 5. 1 The eight symmetry coordinates in terms of the bond stretching coordinates of which the bond polarizabilities are elucidated.

$$S_1 = \frac{1}{\sqrt{2}}(R_{N4C14} + R_{N7C13})$$

$$S_2 = \frac{1}{2}(R_{C5H17} + R_{C3H16} + R_{C8H19} + R_{C12H22})$$

$$S_3 = \frac{1}{2}(R_{C6H18} + R_{C2H15} + R_{C11H21} + R_{C9H20})$$

$$S_4 = R_{C1C10}$$

$$S_5 = \frac{1}{2}(R_{C5C6} + R_{C2C3} + R_{C11C12} + R_{C8C9})$$

$$S_6 = \frac{1}{2}(R_{C1C6} + R_{C1C2} + R_{C9C10} + R_{C10C11})$$

$$S_7 = \frac{1}{2}(R_{C3N4} + R_{N4C5} + R_{N7C8} + R_{N7C12})$$

$$S_8 = \frac{1}{\sqrt{6}}(R_{C14H26} + R_{C14H27} + R_{C14H28} + R_{C13H23} + R_{C13H24} + R_{C13H25})$$

Table 5. 2 The observed wavenumbers, fitted wavenumbers, relative intensities (integrated over wavenumber) and potential energy distribution(PED) for the modes of which the intensities are employed for the elucidation of bond(stretch) polarizabilities. For MV solid, the mode intensities are normalized with that at 1057cm^{-1} taken as 100. For the SERS case, the intensity of the mode at 3067 cm^{-1} under −0. 4V is taken as 100. For MV^{2+}, the data are those at −0. 4V. For MV^{+}, the data are those at −0. 8V. For MV^{0}, the data are those at −1. 2V. The PED is from those of the MV solid. S_i's are the symmetry coordinates. The definitions of the symmetry coordinates in terms of the bond stretching coordinates are shown in Table 5. 1. Those S_i's with $i>8$ are due to the bending coordinates. For short, they are not shown here. (See R. E. Hester, S. Suzuki, J. Phys. Chem. 86, 4626, 1982 for their explicit forms.) Voltage is referred to saturated calomel electrode(SCE).

$MVCl_2$ Solid			MV^{2+}			MV^{+}			MV^{0}			Potential Energy Distribution
Raman Shift/cm^{-1}		Relative Raman Intensity	Raman Shift/cm^{-1}		Relative Raman Intensity	Raman Shift/cm^{-1}		Relative Raman Intensity	Raman Shift/cm^{-1}		Relative Raman Intensity	
Exp.	Fitted		Exp.	Fitted		Exp.	Fitted		Exp.	Fitted		
3077	3084	2	3086	3088	24	3086	3090	6	—	3090	—	S_2 (99)
3059	3051	7	3067	3071	100	3058	3061	22	—	3065	—	S_3 (99)
1657	1652	100	1637	1634	47	1631	1629	19	1644	1652	25	S_4 (52), S_6 (28), S_{10} (16)
1554	1547	16	1515	1519	22	1531	1535	32	1528	1525	74	S_7 (63), S_{13} (19), S_8 (7)
1303	1307	76	1294	1293	46	1291	1296	47	1289	1282	23	S_4 (41), S_6 (37), S_{12} (14)
1201	1195	58	1231	1231	16	1236	1231	26	1237	1231	21	S_1 (44), S_{14} (33), S_{11} (9)
844	844	20	839	840	44	—	840	—	—	841	—	S_6 (32), S_5 (27), S_{10} (19), S_7 (14)
657	661	9	—	660	—	—	660	—	—	660	—	S_1 (46), S_{12} (25), S_6 (19)

The criterion for solving phases is that all the eight bond polarizabilities are positive. This results in only one solution set. Shown in Fig. 5. 3 are the bond polarizabilities as the function of time. For a clear presentation, we plot the bond polarizabilities at $t=0$ and after relaxation at $t=9$ ps and the bond electronic densities calculated by DFT(ub3lyp/cc-pvDZ) in Fig. 5. 4. From the figure, it is noted that in the ground state, charges are more aggregated in the molecular ring. The bond polarizability of C1—C10, C1—C6, N4—C14 (and their equivalent bonds, hereafter, this is always true and will not be repeated.) at $t=0$ are larger. Note that C1—C10 is the bond connecting the two rings and N4—C14 is the branch to the peripheral methyl group. This hints that disturbed charges in the Raman process are prone to shifting toward the ring periphery. Besides, the larger bond polarizabilities of C1—C6 and N4—C5, which are connected to C1—C10 and N4—C14, respectively also support this inference. After relaxation at $t=9$ ps, the bond polarizabilities are close to the bond densities in the ground state.

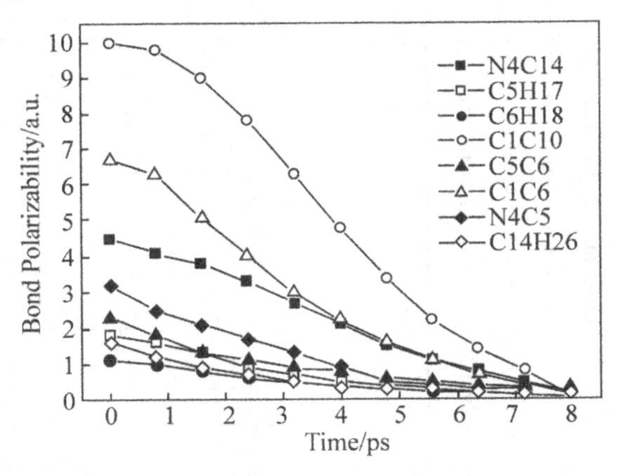

Fig. 5. 3 The bond polarizabilities as the function of time.

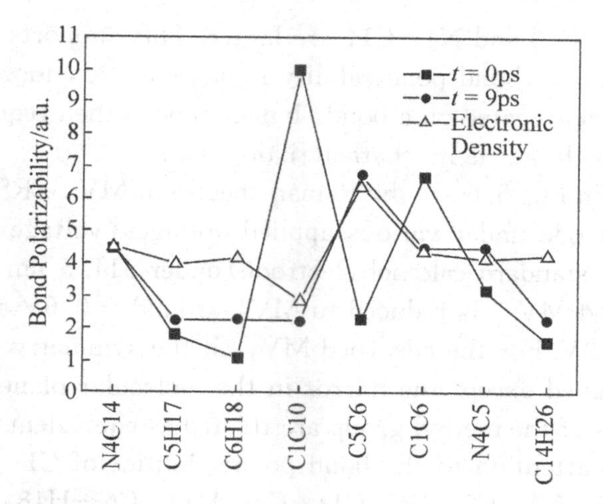

Fig. 5. 4 The bond polarizabilities at $t=0$ and $t=9$ ps and the bond electronic densities.

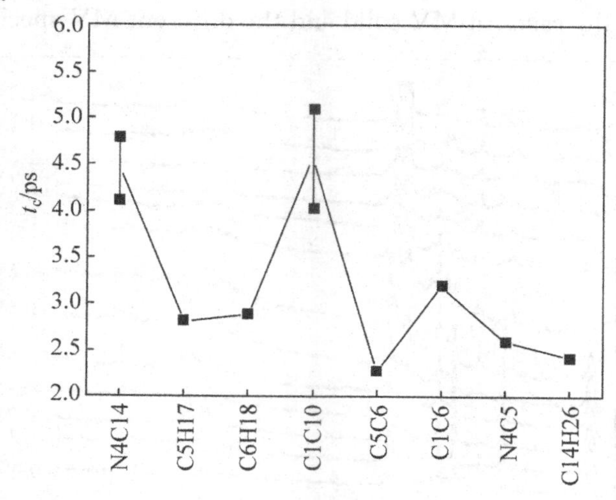

Fig. 5. 5 The characteristic times t_c of the bond polarizabilities.

The relaxation of the bond polarizabilities can be fitted by $y=A\exp(-t/t_c)$, except those of C1—C10 and N4—C14, for which two exponential functions are needed. Shown in Fig. 5. 5 are their characteristic times t_c. From Fig. 5. 5, it is noted that the characteristic times of C1—C10 and N4—C14 are larger. This supports the picture that as the initial bond polarizability is larger, i. e. , more disturbed charges are aggregated on a bond, longer time is then required for its relaxation. So larger is its characteristic time t_c.

Shown in Fig. 5. 6 are the Raman spectra of MV SERS spectra on the Ag electrode under various applied voltages (voltage is with respect to the standard calomel electrode) under 514. 5 nm excitation. The adsorbed MV^{2+} is reduced to MV^{+} around −0. 6V and to MV^{0} around −1. 0V. For the adsorbed MV, all the symmetry elements of D_{2h} are assumed except the mirror in the molecular plane. The three C—H bonds of the methyl group are treated as equivalent. As before, we still pay attention to the bond polarizabilities of C1—C10, C5—C6, C1—C6, N4—C5, N4—C14, C5—H17, C6—H18, C14—H26 (and their equivalent bonds). The normal mode analysis was performed for the cases of MV solid and the different MV species respec-

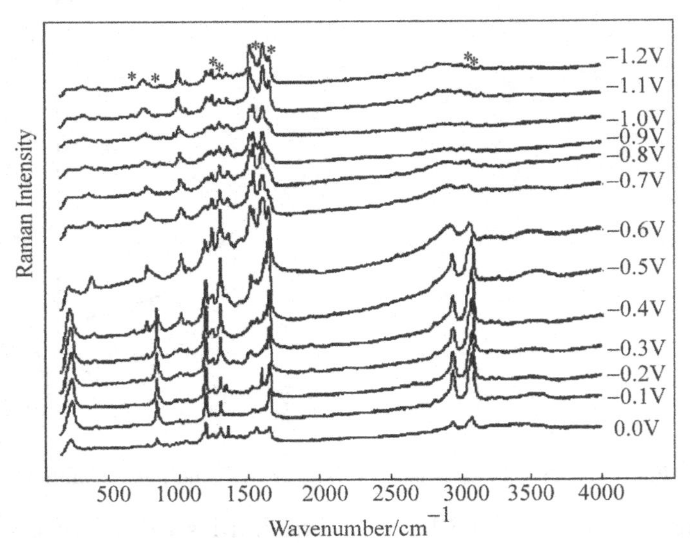

Fig. 5. 6 The Raman spectra of MV SERS spectra on the Ag electrode under various applied voltages by 514. 5 nm excitation. The eight peak intensities employed for the elucidation of bond(stretch) polarizabilities are labeled by *.

tively, since their mode wavenumbers show variation. However, their corresponding $[L_{ij}]$ matrices do not vary significantly, neither do their potential energy distributions (PED). The corresponding observed wavenumbers, fitted wavenumbers, relative intensities (integrated over wavenumber) and PED are tabulated in Table 5. 2. The criterion for solving phases is that all the eight bond polarizabilities are positive. This results in only one solution set for all the MV cases we encountered. Shown in Fig. 5. 7 are the bond polarizabilities of N4—C14, C6—H18 and C5—H17 under various electrode potentials which we will discuss.

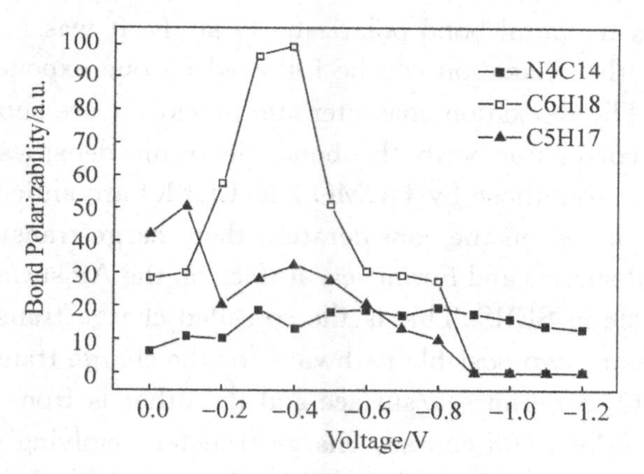

Fig. 5. 7 The bond polarizabilities of N4—C14, C6—H18 and C5—H17 at the very initial moment of Raman excitation under various applied voltages. Note that of C6—H18 at −0. 4 V is normalized to 100.

We have the following observations:

(1) The bond polarizabilities of N4—C14 at various applied potentials do not vary so much as the other ones. This is shown in Fig. 5. 7. We believe this is an indication that N4 and N7 atoms are the adsorption sites. This means that MV molecule is flatly adsorbed on the surface.

(2) At the very initial moment of Raman excitation, another unusual behavior is that the bond polarizabilities of C6—H18 (from 0. 0 V up to −0. 8 V) and C5—H17 (around −0. 1 V) bonds as shown in

Fig. 5. 7 are so large as compared with those of the skeletal bonds. This shows that at the very initial moment of Raman excitation, the excited charges are spread out toward the molecular periphery due to electronic repulsion. We have also observed this behavior in our previous works on 2-, 3-aminopyridine and ethylene thiourea as demonstrated in Chapter 4. We have accessed that this is a generic property of the Raman excited virtual states and this is again confirmed in this MV work.

(3) The relaxation characteristic times of the bond polarizabilities are important parameters of the Raman process from the viewpoint of this temporal bond polarizability study. It was found that in most cases, the relaxation can be followed by one exponential function, e^{-t/t_c}. The relaxation characteristic times, t_c, are shown in Fig. 5. 8. Their correlation with the bond electronic densities (bond orders), especially, those by HOMO and LUMO are anticipated. This viewpoint is based on the consideration that charge transfer between the adsorbed species and Fermi sea/surface on the Ag surface plays an important role in SERS. This is the so-called charge transfer mechanism. There are two possible pathways for the charge transfer: one is from HOMO to Fermi sea/surface and the other is from Fermi sea/surface to LUMO. (Of course, charge transfer involving other occupied and unoccupied MO's is possible. But, the consideration involving HOMO/LUMO is the simplest assumption. We stress that this aspect of charge transfer has nothing to do with the redox of MV. It only involves in the SERS mechanism.) By this conjecture, we infer that the larger a bond electronic density (by HOMO/LUMO) of a bond is, the greater its bond polarizability gains contribution from the charge transfer mechanism and the larger its t_c will be due to that more time is required for the charges to re-distribute during relaxation. The bond electronic densities (bond orders) by HOMO and LUMO, denoted by (…) and <…> respectively, for the various bonds are also shown in Fig. 5. 8. The bond electronic densities by HOMO and LUMO for the C—H bonds are almost zero. Hence, they are not

shown therein. The correlation between large bond electronic density (by HOMO/LUMO)and large relaxation characteristic time is vividly demonstrated. This demonstrates the charge transfer mechanism. Physically, it can be anticipated that the relaxation characteristic time due to the charge transfer mechanism is larger than that due to the electromagnetic mechanism since in the former case, there involves the charge re-distribution and hence longer time is required for the relaxation. Furthermore, the very small bond electronic densities by HOMO/LUMO of these C—H bonds are indeed consistent with their much smaller characteristic times as compared with the cases of the other skeletal bonds. If we consider the bond polarizabilities of the C—H bonds are purely due to the electromagnetic mechanism because of their bare electronic densities by HOMO/LUMO (which implies very little charge transfer mechanism involved), we can figure out from their relaxation characteristic times as shown in Fig. 5. 8 that the time division between the charge transfer and electromagnetic mechanisms is around 3 ps. Of course, we do not rule out the electromagnetic mechanism for those cases with characteristic times larger than 3 ps which are in general associated with larger bond electronic densities by HOMO/LUMO and hence, evidently involve more charge transfer mechanism. Furthermore, from Fig. 5. 8, it is clear that for the three species, MV^{2+}, MV^{+} and MV^{0} (their voltage regions are partitioned by the vertical dotted lines), the relaxation characteristic times of their bond polarizabilities do show variations. The three MV species do show different properties in this aspect. Finally, we note that there is clue that adsorption enhances relaxation, leading to smaller t_c. This is mostly evidenced in the cases of N4—C14, C1—C10 and C1—C6 bonds that their relaxation characteristic times in SERS(at 0. 0 V)are smaller than those in the MV solid.

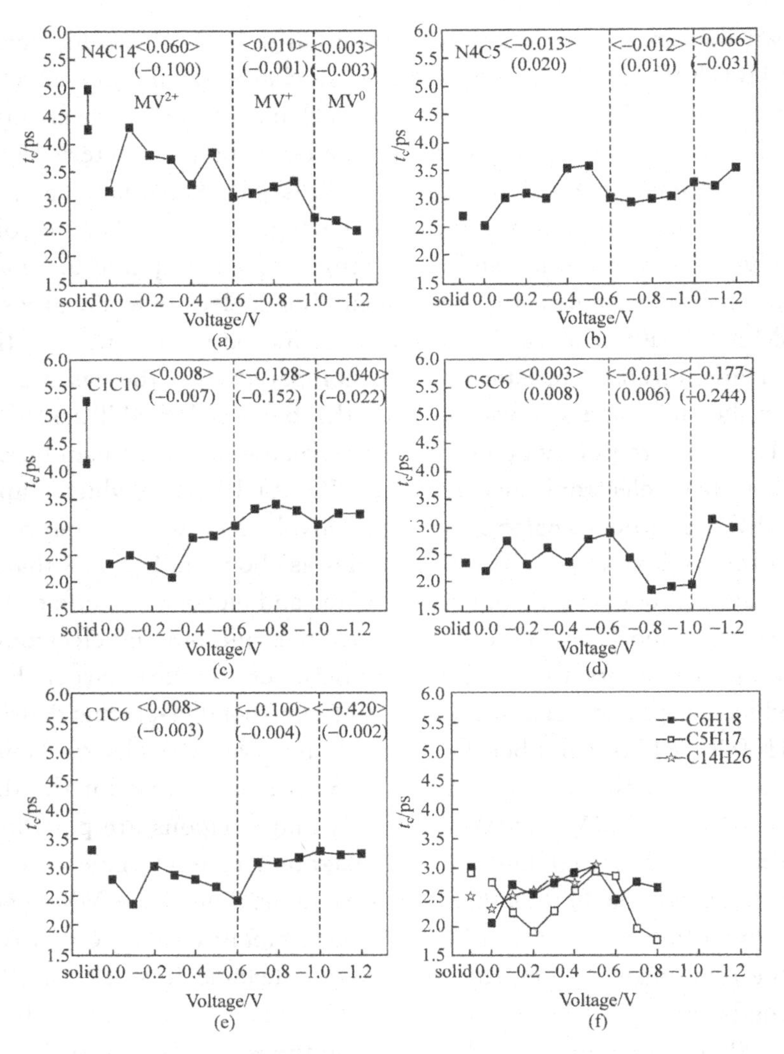

Fig. 5. 8 The relaxation characteristic times, t_c of the temporal bond polarizabilities under various applied voltages and the bond electronic densities (bond orders) by HOMO and LUMO, denoted by(…)and <…> respectively. The dotted lines show the voltage regions for MV^{2+}, MV^{+} and MV^{0}. Note the variations of the characteristic times of the three MV species. Also shown are those of the MV solid for which there are cases apparently with two-stage relaxation and hence with two relaxation characteristic times.

(4) The bond polarizabilities close to complete relaxation which is around 8ps after excitation are of interest. For this, we note first that the bond polarizabilities for the MV solid are very parallel to the bond electronic densities(bond orders) calculated from all the occupied MO as shown in Fig. 5. 9(a). This is very interesting since we can map out the bond electronic densities through the bond polarizabilities near the final stage of their relaxation. This has also been observed in our previous works on ethylene thiourea and 2-, 3-aminopyridine (see Chapter 4). For this case of MV adsorbed on the Ag surface, the bond polarizabilities near the final stage of relaxation are shown in Figs. 5. 9 (b)~(d). From them, we see that though there are variations of the bond polarizabilities among MV^{2+}, MV^{+} and MV^{0}, that of N4—C5 is always the largest. While for the solid case, the bond polarizability of C5—C6 is the largest. We believe that this shows the adsorption effect on N4 and N7 atoms.

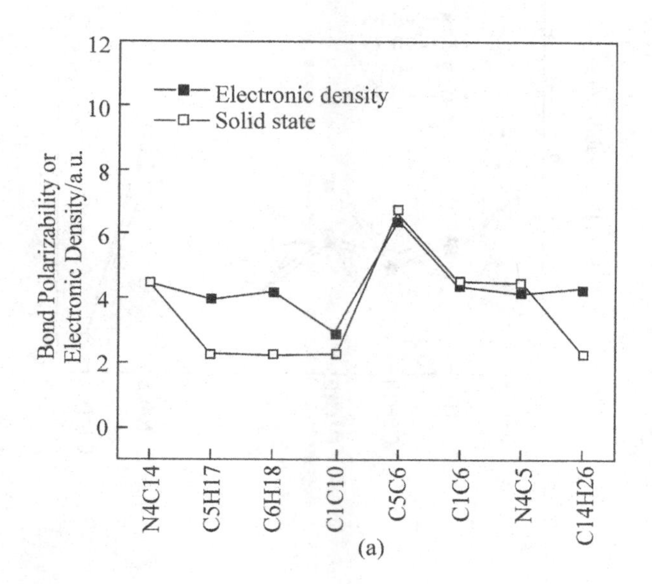

(a)

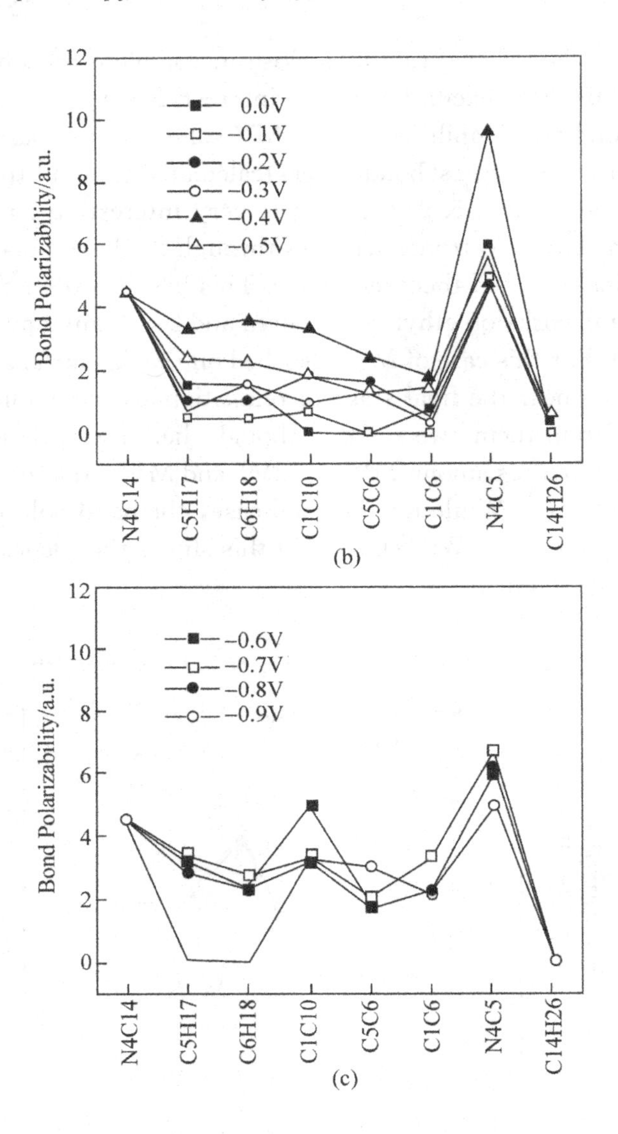
12
10
8
6
4
2
0
Bond Polarizability/a.u.
0.0V
−0.1V
−0.2V
−0.3V
−0.4V
−0.5V
N4C14
C5H17
C6H18
C1C10
C5C6
C1C6
N4C5
C14H26

(b)

12
10
8
6
4
2
0
Bond Polarizability/a.u.
−0.6V
−0.7V
−0.8V
−0.9V
N4C14
C5H17
C6H18
C1C10
C5C6
C1C6
N4C5
C14H26

(c)

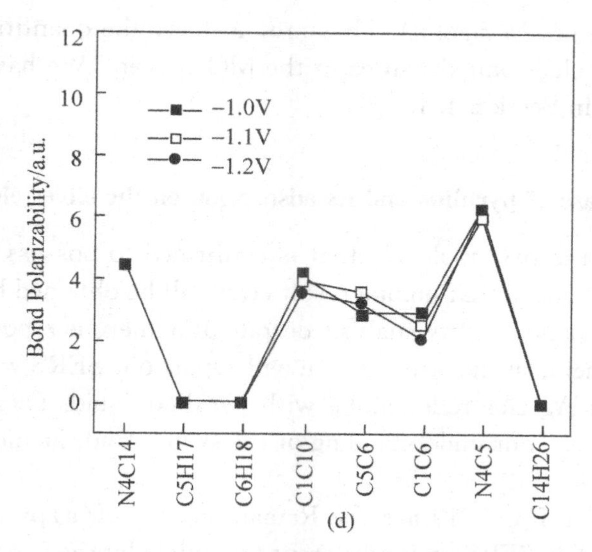

Fig. 5. 9 (a) The bond polarizabilities close to the final stage of relaxation (□) (8 ps after excitation) and the bond electronic densities (bond orders) (■) calculated from all the occupied MO's for the MV solid. (b)～(d), The bond polarizabilities close to the final stage of relaxation for the adsorbed MV under various applied voltages. Note that all the data are normalized with respect to those of the N4—C14 bond.

Comments

(1) The relaxations of the temporal bond polarizabilities show the charge transfer and electromagnetic mechanisms in SERS in this MV case. These two mechanisms involve different relaxation times. The physics is that the process involving in the charge transfer mechanism will take longer time since it needs more time to re-distribute the charges during relaxation. We have observed this phenomenon in the case of ethylene thiourea adsorbed on the silver electrode. See Section 4. 4. This time division for the charge transfer and electromagnetic mechanisms is estimated to be around 3 ps for this MV system.

(2) The adsorption effect can be mapped out from the temporal bond polarizabilities close to the final stage of relaxation (as contrasted

with the case of MV solid). They are, in fact, the quantities parallel to the bond electronic densities in the MO concept. We have also observed this in Section 4. 4.

5. 2 The case of pyridine and its adsorption on the silver electrode

Pyridine is the first molecule that is confirmed to possess SERS[1-5]. We will see indeed that unique results can still be obtained by our proposed bond polarizability analysis despite of numerous reports on it up to this moment. In this section, we will report our SERS work of pyridine on the Ag electrode, along with pyridine liquid. Their comparison will deepen our understanding of the system and the merits of this algorithm.

Shown in Fig. 5. 10 are the Raman spectra of (a) pyridine liquid and (b) pyridine SERS under different potentials by the 514. 5 nm excitation.

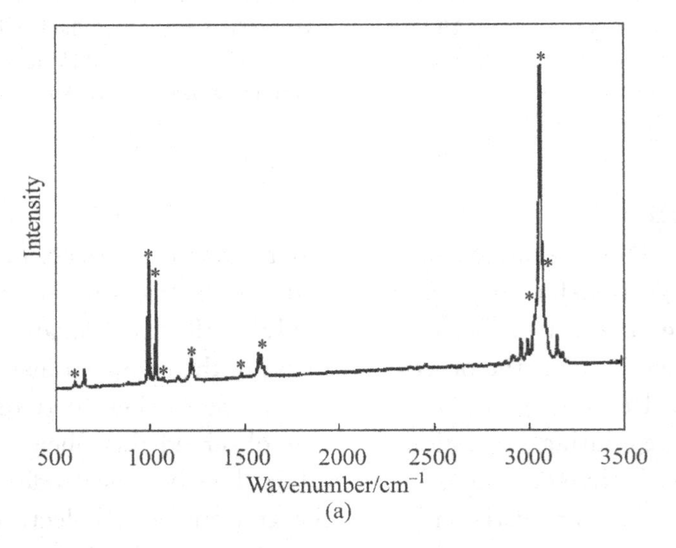

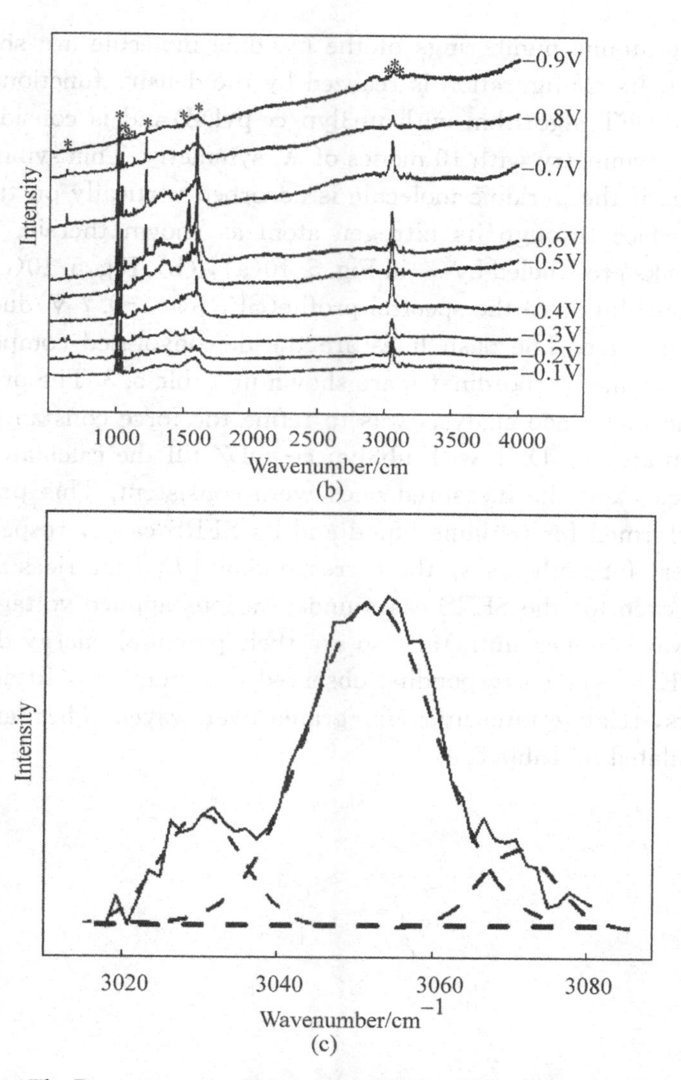

Fig. 5. 10 The Raman spectra of (a) pyridine liquid and (b) pyridine SERS under different potentials by the 514. 5 nm excitation. The symbol * in (a) and (b) shows those peaks that are employed for the elucidation of the bond polarizabilities. (c) shows the deconvolution of the spectral profile (SERS at −0. 7 V) due to the C—H vibration. The dash lines are the de-convoluted components.

The atomic numberings of the pyridine molecule are shown in Fig. 5. 11. Its configuration is realized by the density functional optimization(DFT algorithm with ub3lyp/cc-pvDZ) and is considered to be of C_{2v} symmetry with 10 modes of A_1 symmetry. This symmetry is preserved if the pyridine molecule is adsorbed vertically on the electrode surface through its nitrogen atom as shown therein. The 10 mode peaks are labeled by * in Fig. 5. 10(a), (b). Fig. 5. 10(c) shows the deconvolution of the spectral profile(SERS at −0. 7 V) due to the C—H vibration. The dash lines are the de-convoluted components. The 10 symmetry coordinates are shown in Table 5. 3. The procedure of the normal mode analysis was to refine the force constants which were initiated by DFT with ub3lyp/cc-pvDZ till the calculated mode frequencies and the measured ones were consistent. This procedure was performed for pyridine liquid and its SERS cases, respectively. However, for both cases, the corresponding $[L_{tj}]$ matrices are very similar(even for the SERS cases under various applied voltages with minor wavenumber shifts) and so are their potential energy distributions(PED). The corresponding observed wavenumbers, fitted wavenumbers, relative intensities(integrated over wavenumber) and PED are tabulated in Table 5. 4.

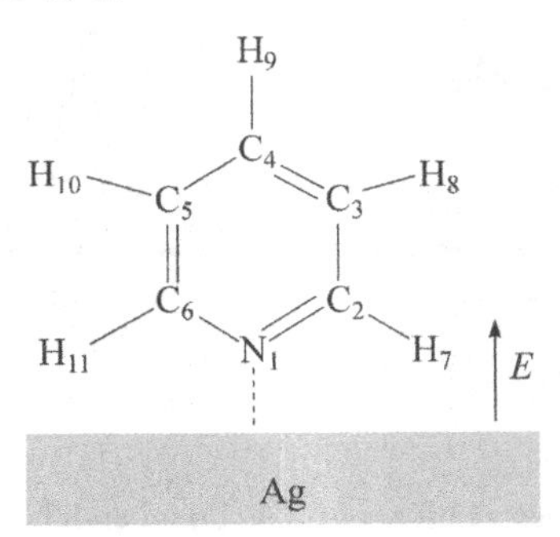

Fig. 5. 11 The atomic numberings of the pyridine molecule. Also shown is its presumed adsorption configuration on the Ag electrode. The arrow shows the direction of the electric field operating in the SERS electromagnetic mechanism.

Table 5. 3 The A_1 symmetry coordinates of pyridine in terms of the bond stretching and angular coordinates. $\alpha_{i,j,k}$ shows the angular deformation defined by the atoms i, j and k. See Fig. 5. 11 for the atomic numberings.

$$S_1 = \frac{1}{\sqrt{2}}(\mathrm{N1C2} + \mathrm{N1C6})$$

$$S_2 = \frac{1}{\sqrt{2}}(\mathrm{C2C3} + \mathrm{C5C6})$$

$$S_3 = \frac{1}{\sqrt{2}}(\mathrm{C3C4} + \mathrm{C4C5})$$

$$S_4 = \frac{1}{\sqrt{2}}(\mathrm{C2H7} + \mathrm{C6H11})$$

$$S_5 = \frac{1}{\sqrt{2}}(\mathrm{C3H8} + \mathrm{C5H10})$$

$$S_6 = \mathrm{C4H9}$$

$$S_7 = \frac{1}{\sqrt{6}}(\alpha_{6,1,2} - \alpha_{1,2,3} + \alpha_{2,3,4} - \alpha_{3,4,5} + \alpha_{6,5,4} - \alpha_{1,6,5})$$

$$S_8 = \frac{1}{2\sqrt{3}}(2\alpha_{6,1,2} - \alpha_{1,2,3} - \alpha_{2,3,4} + 2\alpha_{3,4,5} - \alpha_{6,5,4} - \alpha_{1,6,5})$$

$$S_9 = \frac{1}{2}(\alpha_{1,2,7} - \alpha_{3,2,7} + \alpha_{1,6,11} - \alpha_{5,6,11})$$

$$S_{10} = \frac{1}{2}(\alpha_{2,3,8} - \alpha_{4,3,8} + \alpha_{6,5,11} - \alpha_{4,5,10})$$

Table 5. 4 The experimental Raman wavenumbers, the fitted wavenumbers, their relative intensities and the potential energy distributions(only those larger than 10 are shown) for pyridine liquid(the upper ones) and its SERS on the Ag electrode, at −0. 7 V(the lower ones). Since their fitted wavenumbers are the same, they share the same PED and L matrix.

Raman Shift/cm^{-1}			Intensity	Potential Energy Distribution
No.	Exp.	Fitted	514. 5nm	
ν_1	3068	3069	30	S_5 67; S_6 32
	3070		22	

ν_2	3055	3056	100	S_6 63; S_5 30
	3055		88	
ν_3	3030	3030	34	S_4 80; S_5 13;
	3033		34	
ν_4	1583	1582	16	S_2 46; S_3 17; S_{10} 13; S_9 11
	1583		100	
ν_5	1482	1480	4	S_9 45; S_{10} 28; S_1 15; S_3 12
	1482		9	
ν_6	1219	1220	20	S_9 44; S_1 23; S_{10} 19; S_2 13
	1213		79	
ν_7	1071	1073	3	S_{10} 40; S_3 20; S_1 19; S_7 15
	1067		15	
ν_8	1029	1027	20	S_7 42; S_3 38; S_2 17
	1029		56	
ν_9	991	997	22	S_1 38; S_7 30; S_2 19; S_3 10
	1002		94	
ν_{10}	603	611	10	S_8 97
	622		26	

The criterion for the phase choice is that: all the bond stretching polarizabilities are positive as time elapses. With this criterion, we found that there are two phase solutions for pyridine liquid and the SERS cases. However, we are very fortunate that among these two solutions, the corresponding bond polarizabilities are very similar. They are shown in Fig. 5. 12(a), (b)(note that the equivalent bonds are not shown for short). The(relative) bond electronic densities of the ground state calculated by DFT with ub3lyp/cc-pvDZ are also shown in Fig. 5. 12(a) for the comparison to the bond polarizabilities after relaxation(roughly about 8 ps after the initial excitation).

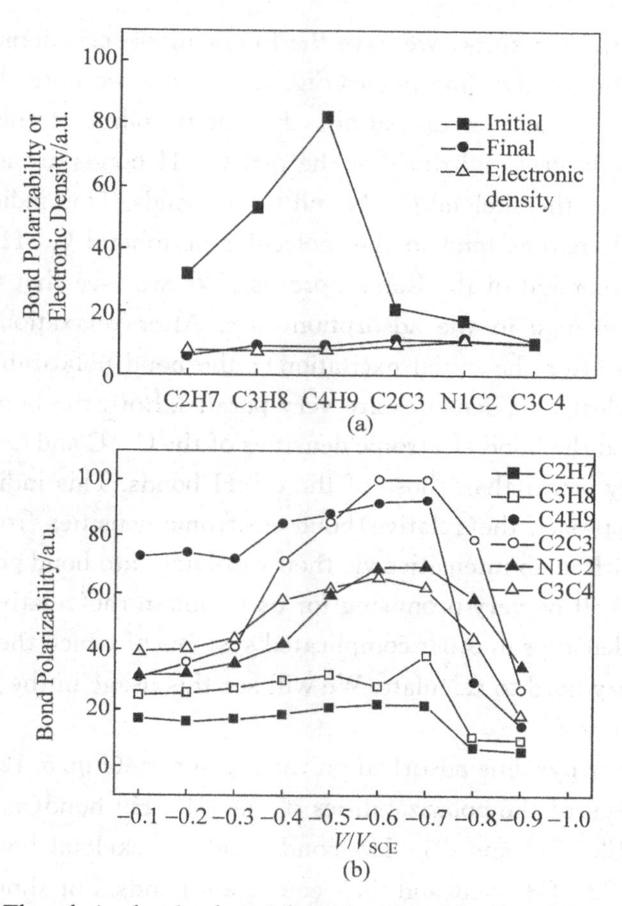

Fig. 5.12 The relative bond polarizabilities for (a) pyridine liquid (both the values at the initial moment (■) of Raman excitation and the final stage (●) of relaxation). Shown therein are also the relative calculated bond electronic densities (△) of the ground state of pyridine molecule. Those of C3—C4 are normalized to 10, for convenience. Note that there is no correlation between the values of the bond polarizability and the bond electronic density. (b) pyridine SERS case (initial moment of Raman excitation). The bond polarizability of C2—C3 at −0.7V is normalized to 100. Note those of the equivalent bonds are not shown for short.

With these results, we have the following observations:

(1) For the pyridine liquid(Fig. 5. 12(a)), we note that at the very initial moment of excitation, the polarizability of the C4—H9 bond is the largest and those of the rest C—H bonds are also larger than those of the skeletal C—N and C—C bonds. This indicates that the excited charges tend to the molecular peripheral C—H bonds at the initial moment of the Raman process. We will see that this behavior does change in the adsorption case. After relaxation (roughly about 8 ps after the initial excitation), the bond polarizabilities and the bond electronic densities are very parallel. Both the bond polarizabilities and the bond electronic densities of the C—C and C—N bonds are slightly larger than those of the C—H bonds. This indicates that we may approach the(relative)bond electronic densities from the experimental Raman intensities via their resolving into bond polarizabilities. This will be very promising for us to obtain the(relative)bond electronic densities in other complicated systems of which these quantities are very hard to calculate. We will see this usage in the pyridazine adsorption case.

(2) For pyridine adsorbed on the Ag surface(Fig. 5. 12(b)), the enhancement of the polarizabilities of the C4—H9 bond(as compared with the C2—H7 and C3—H8 bonds)and the skeletal bonds, especially the C2—C3 bond(and their equivalent bonds. For short, we will not repeat the equivalent bonds hereafter.)during the whole voltage range is obvious. In the voltage range from −0. 1 to −0. 7 V(w. r. t. SCE), the bond polarizability of the C4—H9 bond stays rather constant, though it enhances slightly. Then, it drops instantly as the voltage reaches −0. 8V. We may regard that SERS effect of the C4—H9 bond originates mainly from the electromagnetic mechanism. The argument is that the C4—H9 bond is vertical to the adsorption surface, and hence parallel to the electric field from the metal surface if the vertical adsorption configuration is adopted. By this orientation,

the polarizability of the C4—H9 bond will be more enhanced by the electric field than the rest two C—H bonds just as observed since the latter two C—H bonds are not so parallel to the electric field. Furthermore, one notes that the charge transfer in SERS, if it does occur, will hardly involve the C4—H9 bond (and the rest two C—H bonds) due to its deficiency in the benzenoid π charge which would play the main role for the charge transfer mechanism. As this behavior of the C4—H9 bond is compared with those of the skeletal bonds which enhance significantly as the voltage sweeps from −0.4V to −0.7 V, we assert that, in this voltage range, the polarizabilities of the skeletal bonds are enhanced more through the charge transfer mechanism via the benzenoid π charge. (Of course, their enhancement through the electromagnetic mechanism is not completely ruled out.) We note that the polarizability enhancements of the N1—C2 and C3—C4 bonds are of the same order and less than that of the C2—C3 bond. This hints the conjugation effect if the charge transfer to/from the electrode surface is drawn.

The polarizability enhancements of the C2—H7 and C3—H8 bonds are barely eminent during the whole voltage range. This is not unexpected if their deficiency in the π charge and that they are not so parallel to the electric field from the metal surface are noted.

(3) After relaxation (about 8ps) as shown in Fig. 5. 13, the polarizability of the N1—C2 bond is eminently the largest. As this observation is compared to the (relative) calculated bond electronic densities of the ground state of pyridine molecule (also shown in Fig. 5. 13), we infer that the adsorption site is indeed on the N atom. We note that for the liquid case, the bond polarizabilities after relaxation (shown in Fig. 5. 12 (a)) are very congruent to the bond electronic densities. Therefore, these bond polarizabilities retrieved from the Raman intensities under adsorption, in fact, reflect the adsorption effect.

(4) All the relaxations of the temporal bond polarizabilities

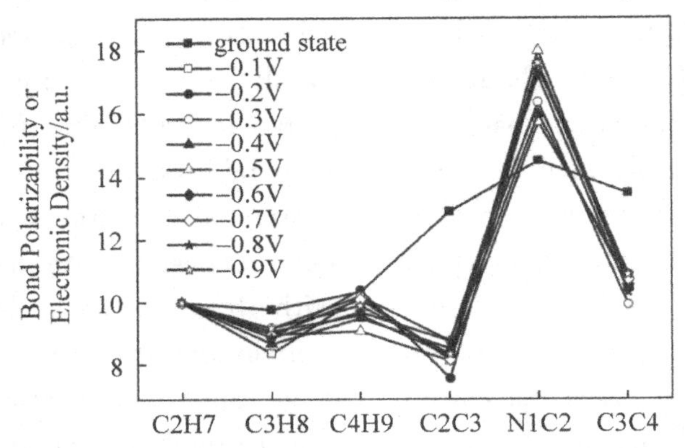

Fig. 5. 13 The relative bond polarizabilities after relaxation (8 ps after initial excitation) for pyridine SERS. Also shown are the relative calculated bond electronic densities of the ground state. The values of C2—H7 under various applied voltages and its bond electronic density are normalized to 10 for convenience. Note that there is no correlation between the values of the bond polarizability and the bond electronic density.

follow an exponentially decaying function. Their relaxation characteristic times, t_c, are shown in Fig. 5. 14 by the fit to the function of $A\exp(-t/t_c)+B$. It is found that the characteristic times of the skeletal bonds are larger than those of the peripheral C—H bonds(under a designated applied voltage). This trend is opposite to the liquid case (also shown in Fig. 5. 14). This can be interpreted as the skeletal C—C and C—N bonds are more involved in the charge transfer mechanism than the peripheral C—H bonds. This is based on the conjecture that the bond polarizability relaxation for the charge transfer mechanism will be longer than that for the electromagnetic mechanism since the former involves the transfer of charges(charge re-distribution)and therefore needs longer time for its relaxation. We note also that as the applied voltage is shifted from −0. 1V toward −0. 7V, the characteristic times become larger for all the bonds in general, and especially for the skeletal bonds. This shows that the charge transfer mechanism becomes more eminent as the applied voltage is shifted toward −0.7 V and that this mechanism is more involved with the skeletal bonds as discussed previously.

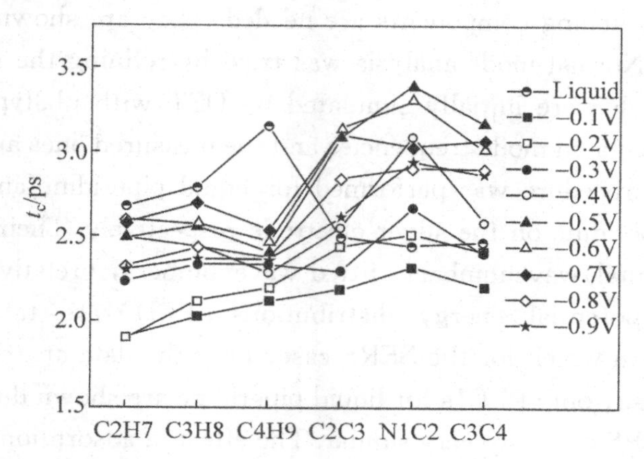

Fig. 5.14 The relaxation characteristic times, t_c, for the bond polarizabilities of pyridine adsorbed on the Ag electrode under various applied voltages and the liquid case.

Comments

The SERS and liquid cases of pyridine are typical Raman topics in numerous studies. Our algorithm of retrieving bond polarizabilities from Raman intensities does offer us some results that have never been known before. This demonstrates the adequacy and power of this algorithm.

5.3 The case of piperidine and its adsorption on the silver electrode

Piperidine is a SERS sensitive compound. Shown in Fig. 5.15 is its structure and bond stretching coordinates. Shown in Fig. 5.16 are the Raman spectra of liquid piperidine and its SERS spectra under various potentials by 514.5 nm excitation. The configuration of piperidine is of C_s symmetry. There are 25 symmetric modes among its 45 normal modes. We will focus on the 10 bond stretching polarizabilities. (The corresponding symmetry coordinates adopted are shown in Table 5. 5.) Therefore, only 10 Raman peaks whose modes are mostly of bond stretching components are needed. They are shown by * in Fig. 5. 16. Normal mode analysis was tried by refining the force constants which were initially simulated by DFT with ub3lyp/cc-pvDZ till the calculated mode frequencies and the measured ones are consistent. This procedure was performed for liquid piperidine and the absorbed piperidine on the silver electrode respectively. Their observed (experimental) wavenumbers, fitted wavenumbers, (relative) intensities and potential energy distributions (PED) are tabulated in Table 5. 6 in which for the SERS case, only the data at −1. 0 V are shown. Also, only PED's for liquid piperidine are shown due to those for the SERS cases are very similar. The effect of adsorption under various applied voltages on $[L_{tj}]$ matrix is negligible so that one may use the same matrix for all cases.

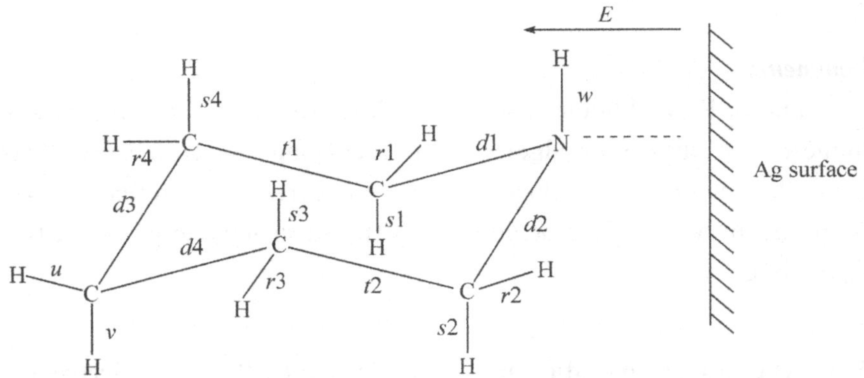

Fig. 5. 15 The molecular structure of piperidine and its bond stretching coordinates. The proposed adsorption configuration on the Ag electrode surface and the direction of the electric field originating from the surface.

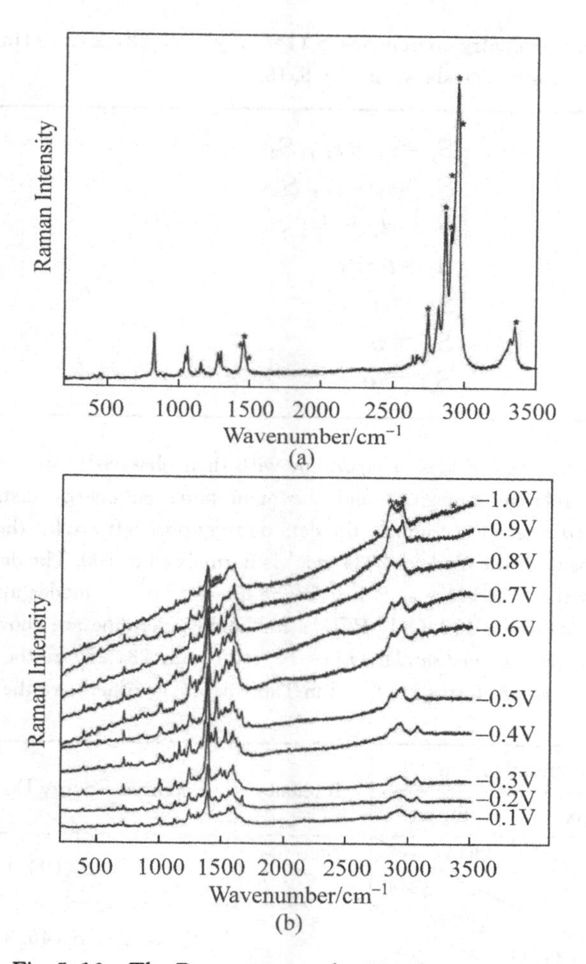

Fig. 5. 16 The Raman spectra by 514. 5 nm excitation of (a) liquid piperidine and (b) its SERS under various potentials. The 10 Raman peaks whose modes are mostly of bond stretching components are labelled by $*$. Their wavenumbers are listed in Table 5. 6.

Table 5.5 The symmetry coordinates S_i ($i=1$, …, 7, 22, 23, 25) in terms of the bond stretching coordinates shown in Fig. 5.15.

$$S_1=r_1+r_2, \quad S_2=r_3+r_4$$
$$S_3=s_1+s_2, \quad S_4=s_3+s_4$$
$$S_5=d_1+d_2, \quad S_6=d_3+d_4$$
$$S_7=t_1+t_2$$
$$S_{22}=u$$
$$S_{23}=v$$
$$S_{25}=w$$

Table 5.6 The normal modes of piperidine with their observed (experimental), fitted wavenumbers, relative intensities and the main potential energy distributions. For wavenumber and intensity columns, the data on the upper left are for the liquid piperidine. The intensity of the mode at 2934 cm^{-1} is normalized to 100. The data on the lower right are for the SERS case at −1.0V. The intensity of the modes at 2842 cm^{-1} at −1.0V is normalized to 100. Only PED's for liquid piperidine are shown since those for the SERS cases are very similar. S_i ($i=1$, …, 7, 22, 23, 25) are the symmetry coordinates. Their explicit forms are listed in Table 5.5. The others are the bending coordinates.

Mode	Wavenumber/cm^{-1}		Intensity	Potential Energy Distribution/%
	Exp.	Fitted		
ν_1	3342 / —	3345 / 3344	53 / 0	S_{25}(99.1)
ν_2	2947 / 2941	2950 / 2943	76 / 92	S_2(47.2), S_4(46.3), S_{22}(5.1)
ν_3	2934 / 2924	2930 / 2928	100 / 73	S_{22}(50.4), S_{23}(47.3)
ν_4	2918 / 2912	2921 / 2917	69 / 67	S_1(92.1), S_3(6.4), S_7(1.3)
ν_5	2895 / 2886	2889 / 2889	49 / 87	S_4(50.0), S_2(48.9)

ν_6	2853 / 2842	2850 / 2850	72 / 100	S_{23}(49.3), S_{22}(48.6)
ν_7	2731 / —	2730 / 2731	42 / 0	S_3(93.4), S_1(3.1), S_7(2.1),
ν_8	1474 / —	1474 / 1476	12 / 0	S_7(41.3), S_5(30.2), S_{18}(14.0), S_{12}(10.9), S_{24}(2.7)
ν_9	1450 / 1455	1453 / 1453	34 / 1	S_{21}(80.1), S_7(9.4), S_{15}(8.2),
ν_{10}	1441 / —	1440 / 1441	25 / 0	S_5(50.3), S_{10}(21.4), S_{12}(16.8), S_{15}(10.3)

The criterion for the phase choice is that: all the bond polarizabilities are of the same sign as time elapses. With this criterion, we were left with only four phase solutions. However, these four solutions are very similar and one of them is chosen for demonstration. All the bond polarizabilities follow the monotonically decaying behavior as time elapses. The (relative) bond polarizabilities of u, v, w, $r1$, $r3$, $s1$, $s3$, $d1$, $d3$, $t1$ coordinates (Their equivalents under C_s symmetry are not shown for short.) are shown in Fig. 5. 17 and Fig. 5. 18 for the liquid and SERS cases respectively. The bond electronic densities of the ground state calculated by the quantal method are also shown in Fig. 5. 17 for comparison. The relaxation characteristic times of the bond polarizabilities are important parameters from the viewpoint of this temporal bond polarizability study. It was found in most cases the relaxation can be well followed by one exponential function of the form: $Ae^{-t/t_c}+B$. The relaxation characteristic times, t_c, for all the cases are shown in Fig. 5. 19.

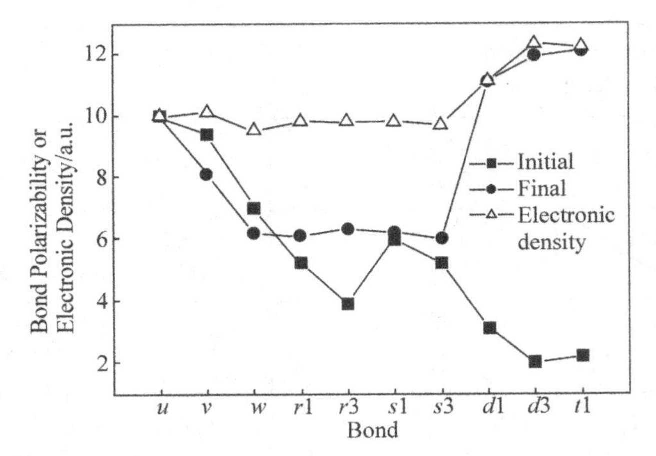

Fig. 5. 17 The(relative)bond polarizabilities of u, v, w, $r1$, $r3$, $s1$, $s3$, $d1$, $d3$, $t1$ coordinates (for their definitions, see Fig. 5. 15)for liquid piperidine at the initial(■) and final(●) stages of Raman relaxation. Also shown are the quantum mechanically calculated bond electronic densities(△) of the ground state. All data are normalized with respect to those of u bond taken as 10. It should be noted that there is no correlation between the values of the bond polarizability and the calculated bond electronic density.

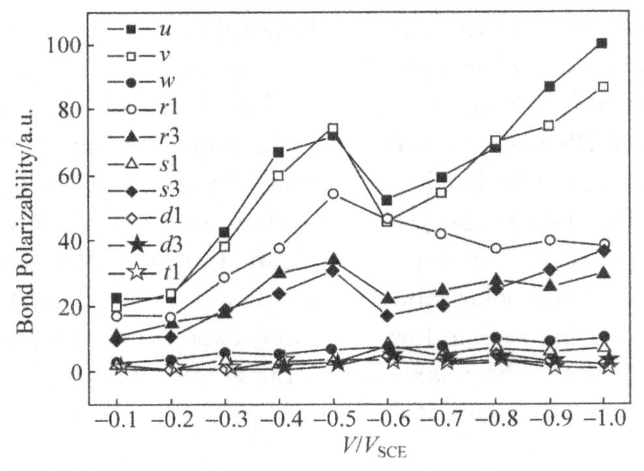

Fig. 5. 18 The(relative)bond polarizabilities of u, v, w, $r1$, $r3$, $s1$, $s3$, $d1$, $d3$, $t1$ coordinates at the initial stage of Raman process for piperidine SERS cases under various applied voltages. All data are normalized with respect to that of u bond at -1.0 V taken as 100.

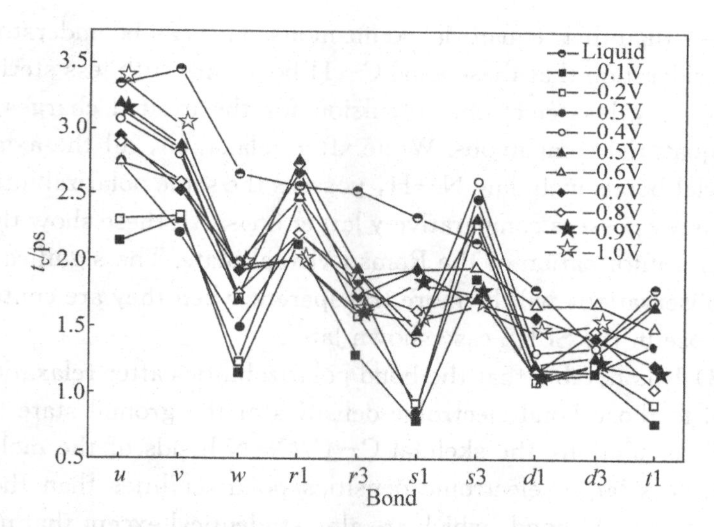

Fig. 5. 19 The relaxation characteristic times, t_c, of various bonds for the liquid and SERS piperidine cases.

A. For the liquidpiperidine case, we have the following observations:

(1) At the initial stage of Raman excitation, the bond polarizabilities of the peripheral C—H/N—H bonds are evidently larger than those of the skeletal C—C/C—N bonds (See Fig. 5. 17). After relaxation(as $t=8$ ps), the skeletal C—C/C—N bonds possess larger bond polarizabilities instead. This shows that at the very initial moment of Raman excitation, the charges are excited in the virtual state and are spread out toward the molecular periphery due to electronic repulsion. We have also observed this behavior in our previous works on 2-, 3-aminopyridine, ethylene thiourea, methylviologen and pyridine. We have accessed that this is a generic property of the Raman excited virtual states and this is again confirmed in this piperidine work.

(2) We further note that at the initial stage of Raman excitation, the axial $s1$ and $s3$ bonds(and their equivalent $s2$ and $s4$. For short, we will not repeat these equivalent bonds hereafter.)possess larger polar-

izabilities than their equatorial companions. This can be understood by the consideration that these axial C—H bonds are with less steric hindrance (i. e., less electronic repulsion for the excited charges) than their equatorial companions. While after relaxation, all the axial and equatorial bonds including N—H, possess the same polarizabilities albeit u and v possess comparatively larger ones. All these show the fine structural information of the Raman virtual state. The significance of these observations will be more transparent when they are contrasted with those in the SERS case shown later.

(3) It is notable that the bond polarizabilities after relaxation are parallel to those bond electronic densities of the ground state by the quantal calculation: the skeletal C—C/C—N bonds of the molecular ring possess larger electronic densities/polarizabilities than those of the C—H/N—H bonds which are almost identical except that u and v bonds possess slightly larger electronic densities/polarizabilities. This is very significant if one notes that the calculated bond electronic density is but a theoretical quantity which would be very hard to observe, if not impossible. We have also observed this behavior in our previous works on 2-, 3-aminopyridine, ethylene thiourea, methylviologen and pyridine and this is confirmed again in this piperidine case. We think our works establish one way to access the bond electronic densities of the ground state simply from the bond polarizabilities after relaxation which are retrieved from the Raman intensities.

B. For the adsorption case of piperidine on the Ag electrode, we note that:

(1) At the initial stage of Raman process, we note that N—H bond, unlike in the liquid case, possesses very small polarizability (See Fig. 5. 18). This may be just caused by adsorption. This suggests that the adsorption site is on the N—H moiety. The skeletal C—C/C—N bonds also possess bare polarizabilities during the whole SERS process, probably due to the same cause. We further note that the equatorial C—H bonds in most cases (under various applied voltages) do possess larger polarizabilities than their axial companions. (Note

the situation is opposite in the liquid case.) This observation suggests that the adsorption configuration is vertical as depicted in Fig. 5. 15 and the SERS mechanism involved is mainly the electromagnetic one, since in this way, the electric field originating from the electrode surface is parallel to the equatorial C—H bonds (also shown in Fig. 5. 15). Therefore, these bonds are SERS enhanced more. It is obvious in this piperidine SERS case that the main effect is barely due to the charge transfer mechanism. Otherwise, the polarizabilities of the skeletal bonds would be enhanced significantly. This is uncommon in the SERS molecular systems[6] of which the nitrogen-containing cyclic compounds are mostly often via the charge transfer mechanism though the electromagnetic mechanism is not excluded. We have studied piperidine SERS back in our 1987 report (but this work only refers to the Raman process at $t=0$. We will show its results in the next section) where the polarizabilities of the skeletal C—C/C—N bonds were observed being enhanced significantly and both the charge transfer and electromagnetic mechanisms were eminent therein. The variation of these two observations is, for the moment, attributed to the different surface roughening processes. The surface roughening in our 1987 work was with piperidine presence (in the KCl solution), unlike in this work. These two surface treatments could cause different adsorption effects, leading to different SERS mechanisms.

From Fig. 5. 18, we also note that in general the farther the C—H bonds are away from the adsorption site (less binding), the larger their bond polarizabilities are, except that the bond polarizability of $r1$ is exceptionally large, probably due to the significant electromagnetic enhancement near the surface.

(2) The relaxation characteristic times for the liquid and SERS cases show that the adsorption enhances the relaxation of the Raman virtual states (see Fig. 5. 19). This is indicated by the larger characteristic times for the liquid piperidine, in general. This adsorption effect was also observed in our works on methylviologen and pyridine (C2—H7, C3—H8, C4—H9 of Fig. 5. 14). Furthermore, w(N—H) and its

neighboring $s1$(axial C—H bond) bonds possess the smallest relaxation characteristic times, especially when the applied voltage is between −0. 1 V and −0. 3 V. This could be due to that around this voltage range, the adsorption is stronger and that these bonds are closer to the adsorption site. This is consistent with the observation that the polarizability relaxation of $s1$ bond is faster than that of $s3$ also because $s1$ is closer to the adsorption site. Despite of these significant interpretations, however, there is ambiguity why the polarizability relaxation of $r3$ is faster than that of $r1$ in most SERS cases.

Comments

(1) For piperidine adsorbed on the Ag electrode, it is inferred from the elucidated bond polarizabilities that in this case, the electromagnetic mechanism dominates the surface enhanced Raman effect, showing that the equatorial C—H bond is enhanced more than its axial companion.

(2) The study of the relaxation characteristic times for the bond polarizabilities demonstrates that adsorption enhances the relaxation of the Raman virtual state in this piperidine case.

(3) It is noticeable that the bond polarizability at the final stage of relaxation is parallel to the quantum chemically calculated bond electronic density of the ground state in the liquid piperidine case. This is very significant if one notes that the calculated bond electronic density is but a theoretical quantity which would be very hard to observe, if not impossible. We have also observed this behavior in our previous works on 2-, 3-aminopyridine, ethylene thiourea, methylviologen and pyridine. This is confirmed again in this piperidine case. We think our works establish one way to access the bond electronic densities of the ground state simply from the bond polarizabilities which can be traced to the experimental Raman intensities. For this piperidine adsorption case, however, only the polarizabilities of the peripheral C—H bonds are enhanced significantly by the electromagnetic mechanism. The bond polarizabilities after relaxation, therefore,

are not informative enough for us to interpret the bond electronic densities under adsorption effect as we did in the ethylene thiourea, methylviologen and pyridine cases where the charge transfer mechanism is evident.

Back to 1987, we started the work on bond polarizability. Piperidine adsorbed on the silver electrode was our first case study. Shown in Table 5.7 are the results. From the results, we may infer that:

(1) As the voltage is shifted from −0.2V to −0.8V, the bond polarizabilities increase in general. However, their tendencies are different. This shows that as the voltage is more negative, the adsorption is weaker and the bond polarizabilities will be larger due to less binding on the adsorbed piperidine molecule from the electrode surface.

(2) The bond polarizabilities on the carbon ring are much larger than those of the C—H bonds. According to Yoshino and Bernstein[7], both C—C and C—H are single bonds and their polarizabilities should be roughly equal. This strongly suggests that the electronic abundance along the ring skeleton is exceedingly enhanced and this is very probable due to the electronic sharing between the adsorbed piperidine molecule and the electrode surface. This is just the chemical effect due to the charge transfer mechanism. As the voltage shifts from −0.4V to −0.6V, the polarizabilities along the ring are smaller while the others are smaller when the voltage is between −0.2V and −0.4V. This difference shows very delicate behaviors of the adsorbed piperidine molecule in SERS.

(3) We note that

$$\partial\alpha/\partial d_3 > \partial\alpha/\partial t_1 > \partial\alpha/\partial d_1$$
$$\partial\alpha/\partial v \geqslant \partial\alpha/\partial s_3 > \partial\alpha/\partial s_1$$

and

$$\partial\alpha/\partial u > \partial\alpha/\partial r_3 > \partial\alpha/\partial r_1.$$

This shows that the adsorption point is on the N atom. As a bond is farther away from the N atom, its polarizability will be larger due to less binding from the adsorption site.

(4) $\partial\alpha/\partial w$ ($\partial\alpha/\partial r_{N-H}$) is very small, showing that electrons around the N—H bond are very scarce due to the charge transfer away to the electrode surface.

(5) We note that

$$\partial\alpha/\partial r_3 \geqslant \partial\alpha/\partial s_3$$
$$\partial\alpha/\partial u > \partial\alpha/\partial v$$
$$\partial\alpha/\partial r_1 \geqslant \partial\alpha/\partial s_1.$$

In other words, the equatorial C—H possesses larger polarizability than the axial one does. (At −0. 4V, $\partial\alpha/\partial r_1$ and $\partial\alpha/\partial s_1$ are very small and should not be considered as a counter case.) This can be attributed to the electric field vertical to the adsorption surface. As a piperidine molecule is adsorbed on the surface via its N atom, its equatorial C—H bonds are more or less parallel to the electric field. Hence, their electronic behavior will be more seriously disturbed, leading to the larger polarizabilities. While the axial C—H bonds are more or less vertical to the electric field and therefore possess smaller polarizabilities. This is consistent with the prediction by the electromagnetic mechanism.

In conclusion, we have:

(1) From the bond polarizabilities, it can be inferred that piperidine molecule is adsorbed vertically on the surface with the adsorption site at the N atom.

(2) Both the chemical/charge transfer and physical/electromagnetic mechanisms are operative as demonstrated in the bond polarizabilities.

(3) Table 5. 7 shows very delicate behavior of the bond polarizabilities under SERS. A complete SERS model should be able to reproduce these data. In other words, the bond polarizabilities set a standard for the SERS model.

Table 5.7 The bond polarizabilities of piperidine with $\partial\alpha/\partial u = 1$ at -0.4 V.

Bond polarizability	Voltage(V_{SCE})			
	-0.2	-0.4	-0.6	-0.8
$\partial\alpha/\partial u$	0.8	1.0	1.0	1.5
$\partial\alpha/\partial v$	0.3	0.2	0.2	1.2
$\partial\alpha/d_3$	5.1	6.6	5.1	19.0
$\partial\alpha/\partial t_1$	1.2	1.6	1.3	4.9
$\partial\alpha/\partial r_1$	0.3	~0.0	0.3	0.5
$\partial\alpha/\partial r_3$	0.4	0.2	0.4	0.8
$\partial\alpha/\partial s_1$	~0.0	0.1	~0.0	0.2
$\partial\alpha/\partial s_3$	0.3	0.2	0.2	0.5
$\partial\alpha/\partial d_1$	0.7	0.9	0.9	3.2
$\partial\alpha/\partial w$	~0.0	~0.0	~0.1	0.1

Comments

Our 1987 work is more or less primitive. However, it is consistent with our later work as shown previously, especially in the part of electromagnetic enhancement. The merit of the 1987 work is that both the charge transfer and electromagnetic effects are demonstrated vividly. The difference in surface treatments or the instability of surface structure from a microscopic viewpoint can be the cause for the variations between these two experiments with an elapsed 20 years.

5.4 The case of pyridazine and its adsorption on the silver electrode

The structure and the atomic numberings of pyridazine is shown in Fig. 5.20. The SERS spectra under various potentials by 514.5 nm excitation are shown in Fig. 5.21(a). The background is significant, partly due to the SERS character. This inhibits the analysis in the C—H stretching region. Shown in Fig. 5.21(b) is the deconvolution of the spectrum from 1060 cm^{-1} to 1700 cm^{-1} at -0.1 V. The peaks are at 1575cm^{-1}, 1452cm^{-1}, 1388cm^{-1}, 1284cm^{-1}, 1205cm^{-1} and 1058cm^{-1}. The de-convoluted peak widths are somewhat large. Since only relative intensities are concerned, they may not cause too much uncertainty. The adopted symmetry coordinates are shown in Table 5. 8. The criterion for phase determination is that all the bond polarizabilities are positive. By this, we are left with a unique solution as shown in Fig. 5. 22. The bond densities are calculated by DFT(ub3lyp/cc-pvDZ) and are shown in Fig. 5. 23 along with the bond polarizabilities after relaxation(8 ps).

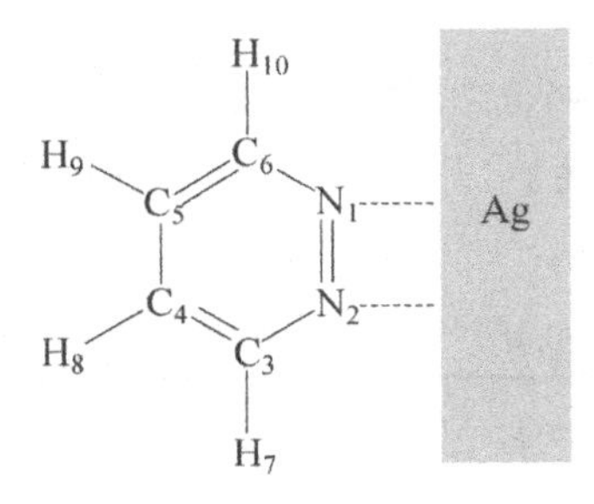

Fig. 5. 20 The structure of pyridazine and its atomic numberings.

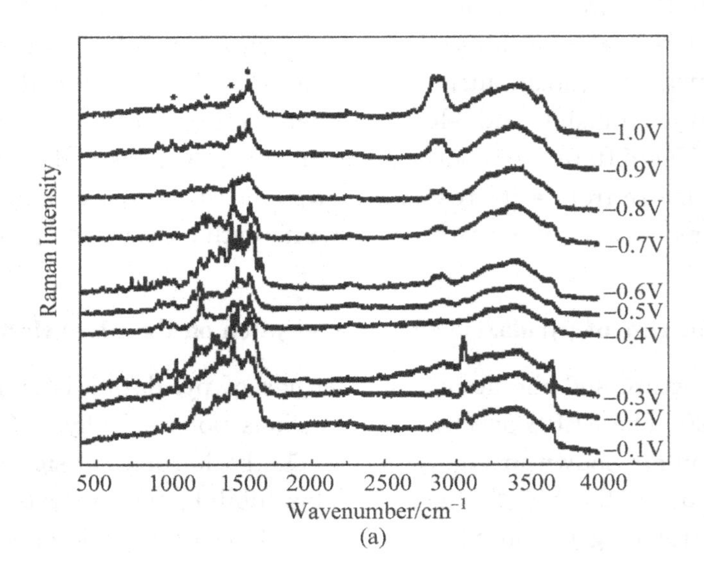

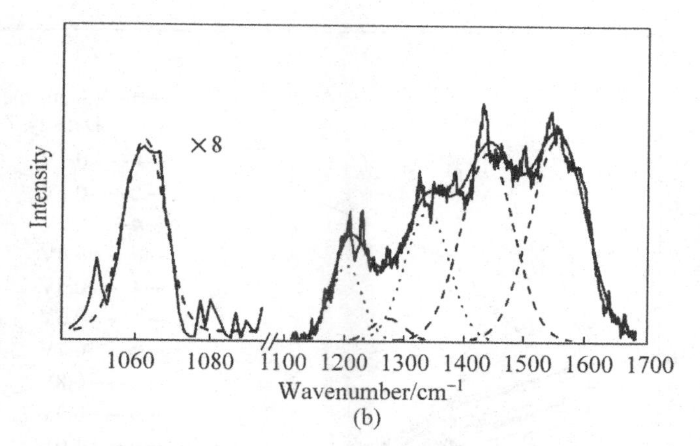

Fig. 5. 21 (a) The SERS spectra of pyridazine on the silver electrode under various potentials by 514. 5 excitation. * are those modes whose intensities are employed for the bond polarizability analysis. (b) The deconvolution of the spectral region at -0.1 V from 1060 cm^{-1} to 1700 cm^{-1}.

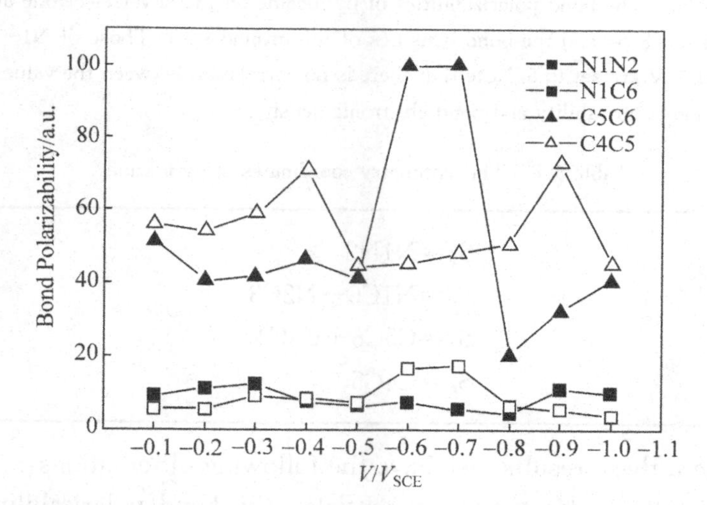

Fig. 5. 22 The bond polarizabilities at the initial Raman excitation. That of C5—C6 at -0.7V is set to 100.

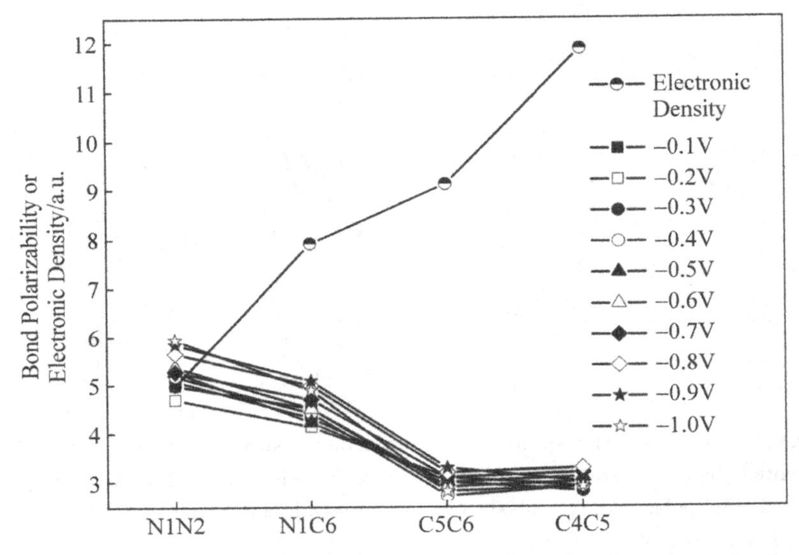

Fig. 5. 23 The bond polarizabilities of pyridazine on the silver electrode after relaxation(8 ps) and the bond densities of the ground state. Those of N1—N2 at −0. 1 V are set to 5. Note that there is no correlation between the values of the bond polarizability and bond electronic density.

Table 5. 8 The symmetry coordinates of pyridazine.

S_1 = N1N2
S_2 = N1C6 + N2C3
S_3 = C5C6 + C3C4
S_4 = C4C5

From these results, we have the following observations:

Overall, under various potentials, the bond polarizabilities of N1—N2 and N1—C6 are smaller and do not change significantly over the potentials. This can be attributed to the strong binding by the adsorption. Those of C5—C6 and C4—C5 are larger and change more significantly over the potentials. The bond polarizabilities can be grouped into two classes. One is N1—N2 and C4—C5, the other is N1—C6 and C5—C6. As the potential shifts from −0.1 V toward −1.0 V, the former two enhance in between −0.1 V and −0.5 V; −0.8V and −1.0 V and decline in between −0.5 V and −0.8V while the latter two show the opposite trend, decline in between −0.1 V and −0.5 V; −0.8V and −1.0 V and enhance in between −0.5 V and −0.8V. If these behaviors are due to the charge transfer mechanism, then the bonding orbital formed by the lone pair electrons of the two N atoms can be the cause of the behaviors of N1—N2 and C4—C5 since it populates more charges on N1—N2 (and hence, through conjugation effect on C4—C5). Similarly, the anti-bonding orbital formed by the lone pair electrons of the two N atoms, which populates more charges on N1—C6(and N2—C3), can be the cause of the behaviors of N1—C6 and C5—C6. The levels of these two orbitals are close to the Fermi surface and may have more chance involving in the SERS mechanism. We will illustrate later our work in 1990 on this system which shows consistent results. There is a difference that here the mode at $1284cm^{-1}$ is assigned to the stretching of N1—N2 and N1—C6 while in the 1990 work, it was the mode at $1388cm^{-1}$. The peaks of these two modes are weak and close to each other. However, the bond polarizability analysis shows consistent results despite of this assignment discrepancy. Here, we appreciate the merits of this algorithm that it is based on a set of mode intensities, instead of one or few ones.

The bond polarizabilities after relaxation(around 8ps) are shown in Fig. 5. 23. Their comparison to the bond densities of the ground state shows the adsorption effect. The bond polarizability of N1—N2 is the largest and that of N1—C6(and N2—C3) the next. This shows that the adsorption is on N1 and N2. Those of C5—C6 and C4—C5 are the least since they are far away from the surface. That the bond polarizabilities after relaxation can offer the adsorption effect is not new to us as being demonstrated in so many previous cases!

The characteristic time t_c can be obtained by the fit with $y = A\exp(-t/t_c) + B$. They are shown in Fig. 5. 24. The characteristic times for N1—N2 and N1—C6 are large, showing that their bond polarizabilities need more time to relax. For C5—C6 and C4—C5, their polarizability behaviors are just in parallel to their bond polarizability behaviors as shown in Fig. 5. 22. In fact, this is also true for N1—N2 and N1—C6, if their trends but not their values are concerned.

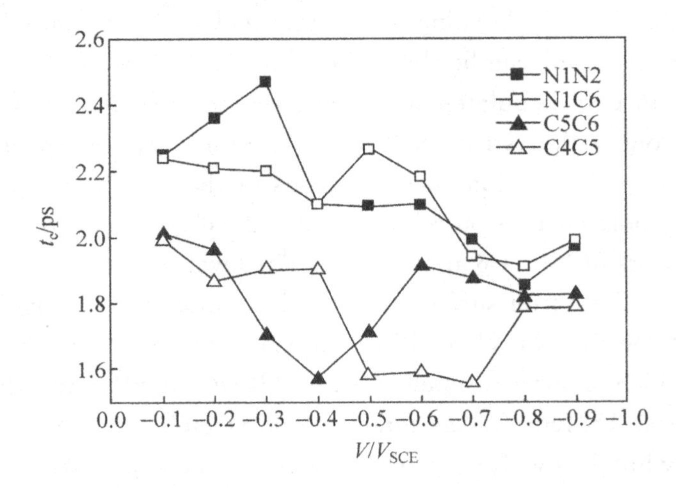

Fig. 5. 24 The characteristic times t_c, of relaxation under various potentials.

Our work in 1990 is shown below. Though its data were not so delicate, valuable results were deduced.

Fig. 5. 25 is the Raman spectrum of pyridazine on the silver electrode. Table 5. 8 shows the symmetry coordinates employed for the bond polarizability analysis. The wavenumbers and relative intensities of the modes after correction are listed in Tables 5. 9 and 5. 10. The four Raman intensities of υ_{8a}, υ_{19b}, υ_{14} and υ_{12} are employed for the calculation of the bond polarizabilities under different applied voltages. The results are shown in Table 5. 11.

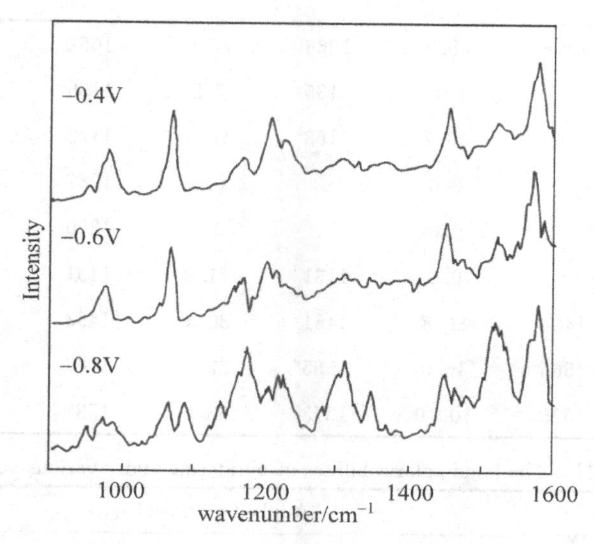

Fig. 5. 25 SERS spectra of pyridazine on the silver electrode.

Table 5. 9 The experimental and the fitted Raman shifts, the intensities and PED.

	Raman Shift/cm^{-1}		Intensity	Potential Energy Distribution
No.	Exp.	Fitted	514. 5nm	
υ_{8a}	1575	1575	100	S_4(70. 3), S_3(10. 2)
υ_{19b}	1452	1460	97	S_3(74. 6), S_4(7. 9)
υ_{14}	1284	1289	10	S_2(70. 7), S_1(6. 9)
υ_{12}	1058	1057	12	S_1(73. 2)

Table 5. 10 The Raman wavenumbers and relative intensities of the modes of pyridazine after correction under various voltages from the silver electrode.

Mode	Applied voltage					
	−0. 4V		−0. 6V		−0. 8V	
	v/cm^{-1}	I	v/cm^{-1}	I	v/cm^{-1}	I
1	976. 5	49. 5	974. 5	47. 3	976	43. 4
12	1055	18. 7	1052	8. 7	1052	20. 7

18b	1068	85. 5	1066	77. 5	1062	30. 8
18a		0. 0	1135	7. 2	1135	26. 5
15	1166	36. 7	1168	16. 6	1172	80. 8
9b		0. 0		0. 0	1287	11. 1
14		0. 0		0. 0	1350	38. 8
19a		0. 0	1131	21. 2	1131	168. 5
19b	1454	80. 8	1451	80. 4	1453	67. 4
8b	1565	10. 0	1565	22. 4	1567	48. 2
8a	1575	100. 0	1573	63. 0	1582	100. 0

Table 5. 11 The bond polarizabilities of pyridazine under various voltages.

Polarizability	Applied voltage(V_{SCE})		
	−0. 4	−0. 6	−0. 8
$\partial\alpha/\partial S_1$	3. 8	2. 3	2. 2
$\partial\alpha/\partial S_2$	−1. 8	−3. 1	−0. 4
$\partial\alpha/\partial S_3$	8. 9	7. 3	7. 1
$\partial\alpha/\partial S_4$	7. 0	5. 7	10. 0

(1) υ_{18a}, υ_{8b} and υ_{19a} show that their normal coordinates are antisymmetric with respect to the mirror bisecting the N—N bond and normal to the molecular plane of pyridazine. In Table 5. 10, the intensities of these modes increase as the voltage shifts from −0. 4V to −0. 8V. This shows that the mirror symmetry is destroyed around −0. 8V since if this symmetry is strict, the intensities of these modes will be very weak or absent.

(2) Shown in Fig. 5. 26 is the very delicate spectral information of υ_1 mode as the voltage shifts from −0. 4V to −0. 8V. As the voltage is from −0. 4V to −0. 6V, its intensity of higher wavenumber portion reduces. While at −0. 8V, the intensity of this portion enhances again. Similar phenomenon also appears in the modes relating to the N—N stretch, such as υ_{8a}, υ_{19b} and υ_{12}. At −0. 8V, the spec-

trum of υ_1 is not so well-defined. One reason is that the adsorption sites are not unique and with a distribution. The following discussion will show that this is related to the charge transfer from the N—N bonding and antibonding orbitals to the electrode surface.

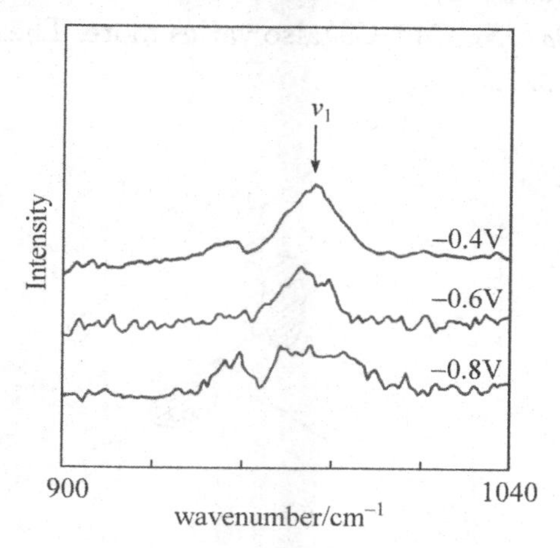

Fig. 5. 26 ν_1 spectrum under different voltages.

(3) As the voltage is close to −0. 8V, except S_4 (C4—C5), all the bond polarizabilities become smaller (see Table 5. 11). At this voltage, the polarizability difference among the bonds is also greater. Then the adsorption is stronger. $\partial\alpha/\partial S_2$ (N1—C6) attains its maximum at −0. 6V, while for $\partial\alpha/\partial S_4$ (C4—C5), it is at −0. 8V. Different bonds behave differently. SERS process is indeed quite complicated.

(4) As the voltage shifts from −0. 4V to −0. 6V, $\partial\alpha/\partial S_1$ (N1—N2) and $\partial\alpha/\partial S_3$ (C5—C6) vary more while as the voltage is from −0. 6V to −0. 8V, $\partial\alpha/\partial S_2$ (N1—C6) and $\partial\alpha/\partial S_4$ (C4—C5) vary more (see Table 5. 11). This can be interpreted in the following way. As the voltage shifts from −0. 4V to −0. 6V, the electrons on the bonding orbital n_1+n_2, formed from the lone pair orbitals n_1 and n_2 of the two N atoms, flow to the electrode. Since the electronic density of n_1+n_2 is concentrated on the N—N bond, the charge transfer will decrease $\partial\alpha/\partial S_1$ (N1—N2). Meanwhile, the electronic shift according to conjugation in the ring will also cause $\partial\alpha/\partial S_3$ (C5—C6) to be smaller. While

as the voltage shifts from −0. 6V to −0. 8V, the charge transfer is from the antibonding orbital $n_1 - n_2$ to the electrode. Since the electronic density of $n_1 - n_2$ is more on the two C—N bonds, the charge transfer will cause $\partial\alpha/\partial S_2$ (N1—C6) to vary more. Meanwhile, by conjugation effect, $\partial\alpha/\partial S_4$ (C4—C5) also varies more. These processes are depicted in Fig. 5. 27.

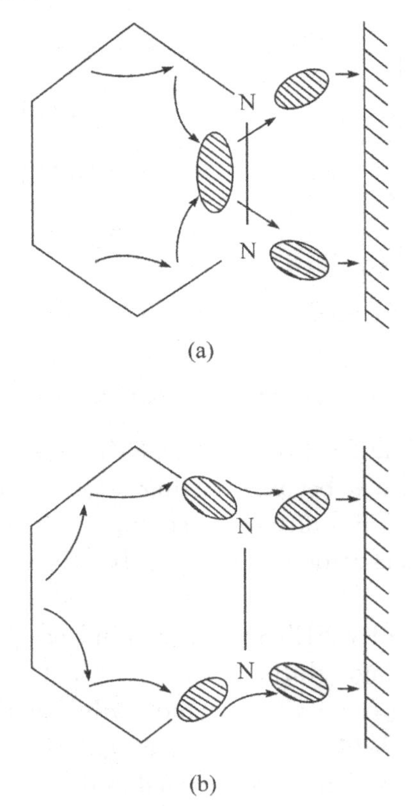

Fig. 5. 27 The charge transfer of (a) bonding orbital $n_1 + n_2$, (b) antibonding orbital $n_1 - n_2$ of pyridazine molecule to the electrode surface.

(5) From the charge transfer model proposed above, it can be inferred that the wavenumbers of those modes involving the N—N stretch will decrease, as the voltage shifts from −0. 4V to −0. 6V,

since then the electronic density on the N—N bonding orbital decreases and hence, the N—N force constant becomes smaller. While as the voltage shifts from −0.6V to −0.8V, the wavenumbers of these modes will increase, since then the electronic density on the N—N antibonding orbital decrease and hence, the N—N force constant becomes larger. As shown in Table 5.10, υ_1, υ_{8a}, υ_{19b}, υ_{12} which involve the N—N stretch do show such a tendency under the applied voltages.

Comments

(1) Evidently, the results of pyridazine share common features with those of piperidine, pyridine and methylviologen. All these demonstrate the common features of the Raman virtual states either of the molecules by themselves or on the electrode.

(2) Though our 1990 work is not so delicate, or even crude, it still shares common features with the later work with an elapsed time of 21 years.

References

[1] Fleischmann M, Hendra P J, McQuillan A J. Chem. Phys. Lett., 1974, 26: 163.
[2] Jeamaire D L, Van Duyne R P. J. Electroanal. Chem., 1977, 84: 1.
[3] Albrecht M G, Creighton J A. J. Am. Chem. Soc., 1977, 99: 5215.
[4] Van Duyne R P. in Chemical and Bio-chemical Applications of Lasers, Vol. 4, C. B. Moore(Ed.), New York: Academic Press, 1979.
[5] Chang R K, Furtak T E. Surface Enhanced Raman Scattering, New York, Plenum Press, 1982.
[6] Otto A. Surface-enhanced Raman Scattering: "Classical" and "Chemical" Origin, in Light Scattering in Solid IV, M. Cardona, G. Guntherodt (Ed), New York: Springer Berlin Heidelberg, 1984: 289—418.
[7] Yoshino T, Bernstein H J. Spectrochim. Acta, 1959, 14: 127.

Chapter 6

The extension to Raman optical activity

6.1 Raman optical activity

Chirality or handness plays an important role in chemistry and biochemistry[1-3]. Its origin stems from the non-superimposability of a molecule on its mirror image. Chiral molecules show variant responses to the right and left circularly polarized light beams. Embedded in these variant responses is stereo-structural information. Spectroscopic techniques such as circular dichroism, vibrational circular dichroism (VCD) and Raman optical activity (ROA) are currently useful for the determination of the absolute configuration of chiral molecules[4-6]. ROA is a measure of the differential Raman cross sections of a chiral molecule by the incident right and left circular polarizations of the exciting laser. The difference of the Raman intensities by these two polarizations is called the ROA spectrum, whose mode intensities are very small, only of the order of 10^{-3} to 10^{-4} of the Raman intensities.

Over the past three decades, ROA has emerged as a powerful spectroscopic technique for the chiral structural determination. It has been proven to be particularly useful in the aqueous environment[7-9] for proteins and nucleic acids. The method by the first principle[10] and the subsequent implementation of ab initio algorithm have resulted in concrete interpretation of the experimental ROA spectra[11-13]. Raman effect involves the vibronic coupling, while ROA effect involves the secondary coupling among the vibrationally induced electric dipole, magnetic dipole and electric quadrupole.

6.2 The differential bond polarizability

For the ROA experiment, we have $I_j^R + I_j^L = I_j$ and $I_j^R - I_j^L = \Delta I_j$ (R and L stand, respectively, for the right and left circularly polarized scatterings and ΔI_j, which is the difference of the Raman intensities by these two scatterings, is called the ROA mode intensity.), i. e.,

$$I_j^R = (I_j + \Delta I_j)/2, \quad I_j^L = (I_j - \Delta I_j)/2.$$

Furthermore, we have

$$\begin{bmatrix} \partial\Delta\alpha/\partial S_1 \\ \partial\Delta\alpha/\partial S_2 \\ \vdots \\ \partial\Delta\alpha/\partial S_t \end{bmatrix} = \begin{bmatrix} & & \\ & a_{jk} & \\ & & \end{bmatrix}^{-1} \begin{bmatrix} P_1(\sqrt{I_1^R} - \sqrt{I_1^L}) \\ P_2(\sqrt{I_2^R} - \sqrt{I_2^L}) \\ \vdots \\ P_t(\sqrt{I_t^R} - \sqrt{I_t^L}) \end{bmatrix}.$$

Here, $\Delta\alpha$ is defined formally as $\alpha^R - \alpha^L$ by $\partial\alpha^R/\partial S_k - \partial\alpha^L/\partial S_k$ which are related, respectively, to the intensities by the right and left circularly polarized scatterings. Since the intensity difference between the right and left circularly polarized scatterings is so small, only of the order of 10^{-3} to 10^{-4} of the Raman intensities, their phases are the same.

Consider

$$\sqrt{2I_j^R} = \sqrt{I_j + \Delta I_j} = \sqrt{I_j}\left[\sqrt{(1 + \Delta I_j/I_j)}\right] \approx \sqrt{I_j}[1 + \Delta I_j/2I_j]$$

$$\sqrt{2I_j^L} = \sqrt{I_j - \Delta I_j} = \sqrt{I_j}\left[\sqrt{(1 - \Delta I_j/I_j)}\right] \approx \sqrt{I_j}[1 - \Delta I_j/2I_j]$$

with error $< [\Delta I_j/I_j]^2/8$. then

$$\sqrt{I_j^R} - \sqrt{I_j^L} \approx \Delta I_j/\sqrt{I_j}$$

and we have

$$\begin{bmatrix} \partial\Delta\alpha/\partial S_1 \\ \partial\Delta\alpha/\partial S_2 \\ \vdots \\ \partial\Delta\alpha/\partial S_t \end{bmatrix} = \begin{bmatrix} & & \\ & a_{jk} & \\ & & \end{bmatrix}^{-1} \begin{bmatrix} P_1(\Delta I_1/\sqrt{I_1}) \\ P_2(\Delta I_2/\sqrt{I_2}) \\ \vdots \\ P_t(\Delta I_t/\sqrt{I_t}) \end{bmatrix}.$$

Though I_j is much larger than ΔI_j by an order of 10^3 to 10^4, the relative magnitudes of I_j's and ΔI_j's can be treated independently as not too small numbers. (For instance, they can be scaled to from 1 to 100 in most cases.) By this way, $\Delta I_j/\sqrt{I_j}$ can be treated as not too small numbers.

In summary, once the bond polarizabilities were obtained from I_j's, then together with the elucidated P_j's and ΔI_j's which are obtained from the ROA experiment, relative $\partial \Delta\alpha / \partial S_k$ can be obtained at hand. $\partial \Delta\alpha / \partial S_k$ can be called the differential bond polarizability, for convenience. This molecular parameter is important for us to understand the electric and magnetic coupling in a ROA molecule. This is impossible by the bond polarizabilities.

Only those portions in a chiral molecule that are close to the asymmetric center contribute significantly to the ROA intensities. The portions which are far away from the asymmetric center contribute much less and hence, their differential bond polarizabilities are small with little ROA information.

Comments

(1) The merit of ROA is that it offers an extra-information in addition to Raman. The extra information is about the molecular stereo- or three dimensional structure. Raman spectrum offers the Raman shifts and mode intensities, which are related to the information about the molecular bonds and barely about the relationship among the bonds, such as their mutual orientations.

(2) Differential bond polarizabilities and $[\Delta I_j / \sqrt{I_j}]$ are related by a linear transformation. Hence they are equivalent. However, differential polarizabilities are obtained from $[\Delta I_j / \sqrt{I_j}]$, by tossing away those factors ($[a_{jk}]^{-1}$) that are not related to Raman/ROA process, i. e., force fields, atomic masses. Also, it is noted that those modes with larger ROA intensities and less Raman intensities contribute more to the differential bond polarizabilities. That is, these are the modes that contain more stereo information. Since differential bond polarizabilities are derived from both ROA and Raman intensities while bond polarizabilities are derived only from Raman intensities, therefore, differential bond polarizabilities bear more information than the bond polarizabilities as far as molecular information is concerned.

(3) People may be inclined to attribute Raman/ROA intensities into ingredients that are induced by various coupling mechanisms. Though this is convenient, we have to admit that this attribution is by arbitration. The observed intensities will contain all the mechanisms whatever they exist. In this sense, bond polarizability and differential bond polarizability which are derived from the Raman/ROA intensities will contain all the mechanism informations as far as they operate in Raman/ROA. Hence, the terminology of polarizability is but a convenient nomenclature. We just cannot misinterpret it.

6.3 The case of (+)-(R)-methyloxirane

Shown in Fig. 6.1 is the structure of methyloxirane and its atomic numberings. Shown in Fig. 6.2 is its Raman spectrum under 514.5 nm excitation. The local symmetry coordinates of(+)-(R)-methyloxirane are defined and described in Table 6.1. Normal mode analysis was run to obtain the L matrix through the fit to the experimental mode wavenumbers by refining the force constants which were initially calculated by DFT with 6-31G* basis.

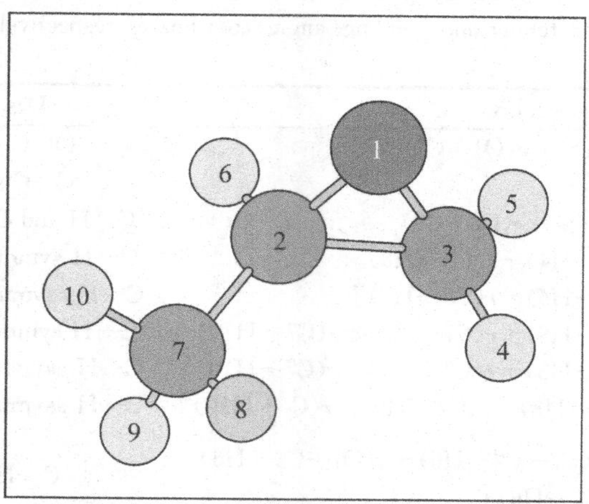

Fig. 6.1 The structure of(+)-(R)-methyloxirane and its atomic numberings. 1 is oxygen atom, 2, 3 and 7are carbon atoms. The rest are hydrogen atoms.

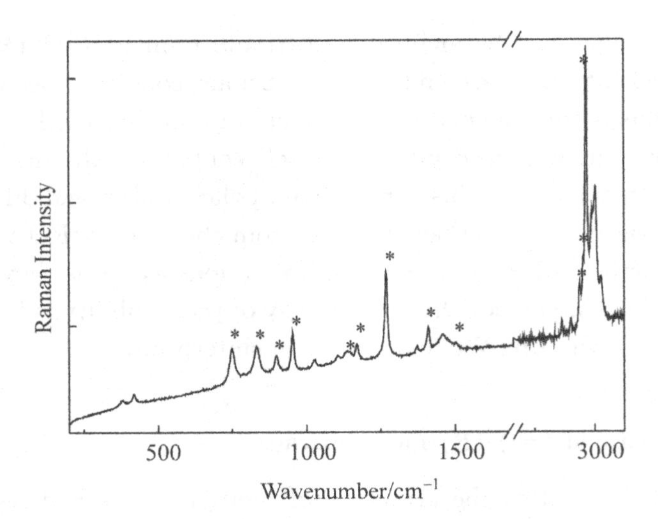

Fig. 6. 2 The Raman spectrum of solid(+)-(R)-methyloxirane by 514. 5 nm excitation. * shows those peak intensities that are employed for the elucidation of the bond stretch polarizabilities. Their Raman shifts and relative intensities are listed in Table 6. 2.

Table 6. 1 The local symmetry coordinates of(+)-(R)-methyloxirane . Here r and a denote the bond stretching and bending(angle) coordinates respectively. τ denotes the dihedral angle.

Definition	Description
$S_1=r(\mathrm{O1-C2})$; $S_2=r(\mathrm{O1-C3})$	O—C stretching
$S_3=r(\mathrm{C2-C3})$	C—C stretching
$S_4=r(\mathrm{C2-H6})$; $S_5=r(\mathrm{C2-C7})$	C—H and C—C stretching
$S_6=2^{-1/2}[r(\mathrm{C3-H4})+r(\mathrm{C3-H5})]$	C—H symmetric stretching
$S_7=2^{-1/2}[r(\mathrm{C3-H4})-r(\mathrm{C3-H5})]$	C—H asymmetric stretching
$S_8=3^{-1/2}[r(\mathrm{C7-H8})+r(\mathrm{C7-H9})+r(\mathrm{C7-H10})]$	C—H symmetric stretching
$S_9=3^{-1/2}[r(\mathrm{C7-H8})+r(\mathrm{C7-H9})-r(\mathrm{C7-H10})]$	C—H asymmetric stretching
$S_{10}=3^{-1/2}[r(\mathrm{C7-H8})-r(\mathrm{C7-H9})+r(\mathrm{C_7-H10})]$	C—H asymmetric stretching
$S_{11}=6^{-1/2}[2\times a(\mathrm{C7-C2-H6})-a(\mathrm{O1-C2-H6})$ $-a(\mathrm{C3-C2-H6})]$	C—H rocking
$S_{12}=2^{-1/2}[a(\mathrm{O1-C2-H6})-a(\mathrm{C3-C2-H6})]$	C—H rocking

$S_{13}=17^{-1/2}[4\times a(\text{O1—C2—C7})+a(\text{C3—C2—C7})]$	C—C—C and C—C—O deformation
$S_{14}=17^{-1/2}[a(\text{O1—C2—C7})+4\times a(\text{C3—C2—C7})]$	C—C—C and C—C—O deformation
$S_{15}=1/2[a(\text{O1—C3—H4})-a(\text{O1—C3—H5})-a(\text{C2—C3—H4})+a(\text{C2—C3—H5})]$	CH_2 twisting
$S_{16}=1/2[a(\text{O1—C3—H4})+a(\text{O1—C3—H5})-a(\text{C2—C3—H4})-a(\text{C2—C3—H5})]$	CH_2 wagging
$S_{17}=1/2[a(\text{O1—C3—H4})-a(\text{O1—C3—H5})+a(\text{C2—C3—H4})-a(\text{C2—C3—H5})]$	CH_2 rocking
$S_{18}=a(\text{H4—C3—H5})$	CH_2 bending
$S_{19}=2^{-1/2}[a(\text{C2—C7—H8})-a(\text{C2—C7—H9})]$	CH_2 rocking
$S_{20}=6^{-1/2}[2a(\text{C2—C7—H10})-a(\text{C2—C7—H8})-a(\text{C2—C7—H9})]$	CH_2 rocking
$S_{21}=2^{-1/2}[a(\text{H9—C7—H10})-a(\text{H8—C7—H10})]$	CH_3 asymmetric deformation
$S_{22}=6^{-1/2}[2a(\text{H8—C7—H9})-a(\text{H9—C7—H10})-a(\text{H8—C7—H10})]$	CH_3 asymmetric deformation
$S_{23}=6^{-1/2}[a(\text{H8—C7—H9})+a(\text{H9—C7—H10})+a(\text{H8—C7—H10})-a(\text{C2—C7—H8})-a(\text{C2—C7—H9})-a(\text{C2—C7—H10})]$	CH_3 symmetric deformation
$S_{24}=a(\text{C3—C2—C7—H8})$	CH out of CCC plane bending

The (+)-(R)-methyloxirane molecule has no symmetry. Its methyl group is out of the plane formed by the triangular oxirane ring. There are 24($3N-6$, $N=10$) vibrational modes for (+)-(R)-methyloxirane. For our purpose of obtaining the bond stretch polarizabilities, as demonstrated below, we will treat those of C3—H4, C3—H5 and C7—8, C7—H9, C7—H10 equivalent respectively. Therefore, we have 7 bond stretch polarizabilities to elucidate. Correspondingly, we need 7 mode intensities whose modes possess most stretching coordinates. Furthermore, we also consider 5 non-negligible intensities whose modes are more of the bending coordinates.

Correspondingly, we have 12 mode intensities to elucidate the 7 bond stretch polarizabilities. They are shown by * in Fig. 6. 2. Their Raman shifts, intensities and potential energy distributions are listed in Table 6. 2. The fit errors of most Raman shifts are less than 1%.

Table 6. 2 The experimental and fitted Raman shifts in cm^{-1}, relative intensities together with their potential energy distributions of (+)-(R)-methyloxirane under 514. 5 nm. The relative mode intensities are with respect to that at 2934 cm^{-1} which is normalized to 100 for convenience.

Experimetnal	Fitted	Intensity	Potential Energy Distribution
2934	2931	100	S_4(97)
2907	2900	20. 7	S_6(99)
2877	2876	12. 3	S_8(85), S_9(4), S_{10}(10)
1503	1504	1. 4	S_{18}(80), S_3(17)
1410	1413	3. 9	S_3(14), S_5(21), S_{14}(27), S_{23}(30)
1268	1265	20. 3	S_2(3), S_3(24), S_{14}(42)
1170	1170	3. 6	S_5(19), S_{12}(16), S_{13}(19), S_{15}(21)
1134	1132	3. 8	S_1(8), S_2(5), S_{11}(8. 9), S_{16}(71)
953	950	8. 7	S_2(38), S_5(26), S_{19}(26)
898	890	5. 7	S_2(13), S_{11}(10), S_{15}(34), S_{17}(20), S_{20}(26)
832	830	16. 6	S_1(37), S_3(31), S_5(10), S_{15}(5)
750	747	16. 5	S_1(45), S_2(36), S_5(11), S_{19}(6)

The criterion for the phase choice is that: all the bond stretch polarizabilities are of the same sign. With this criterion, we are left with 64 phase sets. It was recognized that the relaxation of the bond polarizabilities of the Raman virtual state decays exponentially as time evolves. For these 64 phase sets, if the fluctuations (deviated from the exponential decay) of the bond polarizabilities with time are limited not to be larger than 10% (or 5%) during relaxation, then only 10 solution sets will remain and all possess quite the same bond polarizabilities. The elucidated bond polarizabilities at the initial and final stages (6 ps) are shown in Fig. 6. 3. Shown in Fig. 6. 4 are the bond polarizabilities as a function of time. In fact, this condition of 10% fluctuation is not crucial. All the solutions with this fluctuation condition up to 30% lead to the same observation. The bond polarizabilities of

C2—H6, C2—C3, C2—O1 and C3—O1 are noticeably larger than that of C2—C7, while after relaxation, this trend is opposite. Then, the bond polarizabilities are congruent to the calculated bond electronic densities in the ground state(by RHF/6-31G*) with those of C7—H8 and C3—H4 and their equivalents the least(see Fig. 6. 3). This is the central observation by our algorithm. Hence, our phase solution algorithm is adequate. This also demonstrates that all the electrons in the molecule contribute(with equal probability) to the virtual state. Otherwise, the bond electronic densities will differ from the bond polarizabilities at the final stage. Thus, for the systems like methyloxirane, the bond polarizabilities after relaxation offer us one way to *observe* the *conceptual* bond electronic densities, which are otherwise very difficult to obtain from the experiments, if possible.

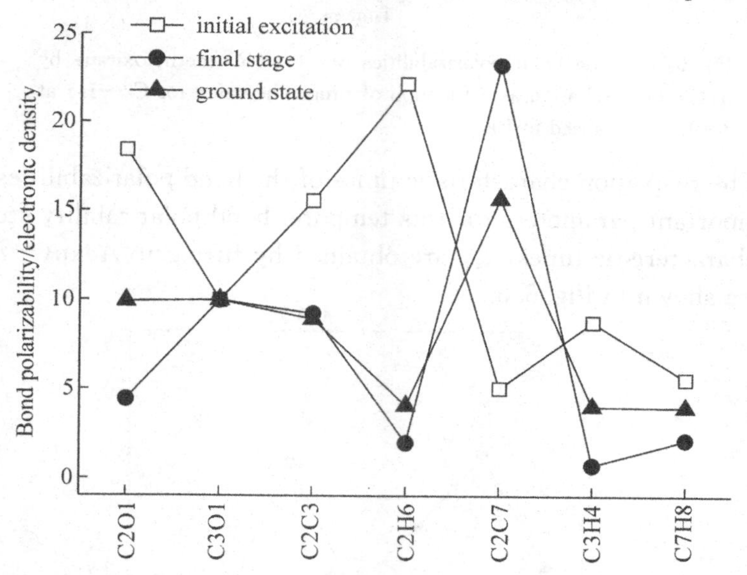

Fig. 6. 3 The relative bond polarizabilities of(+)-(R)-methyloxirane at the initial excitation moment(□) and the final stage of Raman relaxation(●) by 514. 5 nm excitation. (▲) is for the calculated bond electronic densities of the ground state. For convenience, the value of C3—O1 bond is normalized to 10. Note that there is no correlation between the values of the bond polarizability and the calculated bond electronic density.

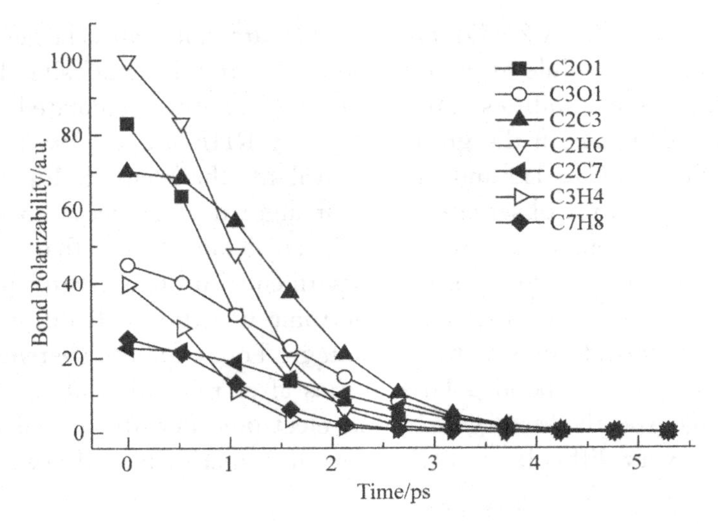

Fig. 6. 4 The bond polarizabilities of (+)-(R)-methyloxirane by 514. 5 nm excitation as a function of time. The value for C2—H6 at $t=0$ is normalized to 100.

The relaxation characteristic times of the bond polarizabilities are the important parameters for this temporal bond polarizability study. The characteristic times, t_c, are obtained by fitting to $A\exp(-t/t_c)$ and are shown in Fig. 6. 5.

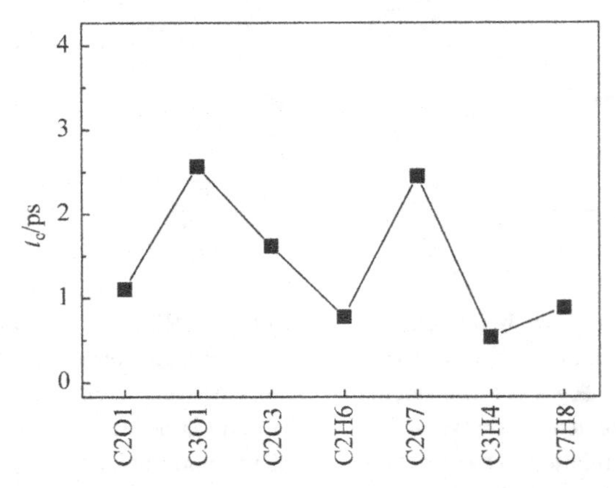

Fig. 6. 5 The characteristic times, t_c, for the relaxations of the various temporal bond polarizabilities of(+)-(R)-methyloxirane by 514. 5 nm excitation.

As shown in Fig. 6. 5, the bond polarizabilities of C3—H4, C2—H6 and C7—H8 (and their equivalents) decay faster than those of the rest bonds. These are the bonds in the molecular periphery. This observation implies that during relaxation, the excited charges around the peripheral bonds will relax first as they are reflowing back toward the molecular core.

As just mentioned, at the initial moment of excitation (as $t=0$ ps), the bond polarizabilities of C2—H6, C2—C3, C2—O1 and C3—O1 are noticeably larger than that of C2—C7. While after relaxation (as $t=6$ ps), this trend is opposite (with those of C7—H8 and C3—H4 and their equivalents the smallest). This demonstrates that initially the tendency of the excited charges is toward the C2—H6 bond and its triangular oxirane skeleton. C2 atom is the asymmetric (chiral) center. The more excited charges distributing on the C2—H6 bond would lead to a significant dipole as it vibrates (as schematically shown in Fig. 6. 6 by the symbol $\boldsymbol{\mu}$). Meanwhile, more excited charges distributing toward the triangular oxirane skeleton near the asymmetric center would induce a significant charge current[14, 15] as it vibrates so that an eminent induced magnetic moment can be expected. (As shown in Fig. 6. 6 by $\boldsymbol{m}$. Note that the Raman relaxation is in the time range of 1 to 10 ps while the bond vibration is in the range of 0. 1 to 0. 01 ps) The coupling of the induced electric dipole and magnetic moment can lead to the significant Raman chirality. (Here, we do not rule out the possible coupling of the electric quadrupole originating from the triangular oxirane skeleton with the induced dipole on the C2—H6 bond though we will not go into its detailed analysis.) With this picture, we expect that the modes with more components of the stretching/bending motion of the triangular oxirane skeleton will possess more significant ROA activity. Indeed, the ROA spectrum shown in Fig. 6. 7[16] does show that the modes around 898 cm^{-1}, 832 cm^{-1} and 750 cm^{-1}, which are mostly of the stretching/bending character of the triangular oxirane as shown by their PED (see Table 6. 2.), possess the most significant ROA activity. Note also that the mode around 953 cm^{-1} does not show significant ROA activity (though quite Raman significant) and it is more of the stretching motion along C2—C7 bond and the bending motion associated with C2—C7—H8(—H9, —H10). Evidently, the Raman chirality of this compound does not come from the dipole on the C2—C7 bond though it possesses the largest bond electronic density in the ground state.

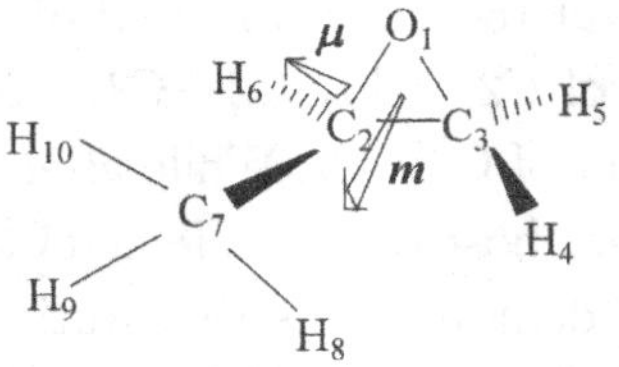

Fig. 6. 6 The sketch of a significant vibrating electronic dipole, $\boldsymbol{\mu}$ induced by the excited charges on the C2—H6 bond and an eminent induced magnetic moment $\boldsymbol{m}$ on the triangular oxirane skeleton near the asymmetric center as the skeleton vibrates in the Raman virtual state. Their coupling results in the significant ROA for the skeletal modes. Note that there is bare excited charges on the C2—C7 bond in the Raman process so that its motion will not contribute to ROA though it possesses the most significant electronic density in the ground state.

Since the current ROA instrument adopts 532 nm excitation, the Raman spectrum of methyloxirane is also taken by 532 nm excitation. Its difference with that by 514. 5 nm is small. Shown in Fig. 6. 8 and Fig. 6. 9 are the spectrum and the elucidated bond polarizabilities by 532 nm excitation. In fact, we have 9 phase sets with rather consistent bond polarizabilities and only one is shown here. Listed in Table 6. 3 are the experimental and fitted Raman shifts in cm^{-1}, relative intensities together with their potential energy distributions. Shown in Fig. 6. 10 is the ROA spectrum which is limited only up to 1600 cm^{-1}. Those due to C—H are around 3000 cm^{-1} and are therefore not available. These C—H bonds are farther away from the asymmetric center, their ROA intensities are assumed to be zero for the moment. Then, we can proceed to elucidate the differential bond polarizabilities. The results are shown in Fig. 6. 11. We note that though we have 9 sets of differential bond polarizabilities, they all show very consistent behaviors.

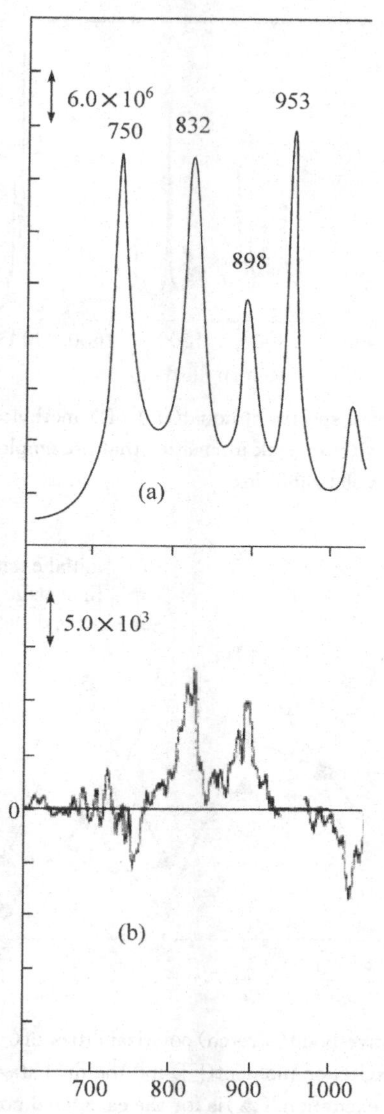

Fig. 6. 7 The Raman (a) and ROA (b) spectra of(+)-(R)-methyloxi ranefrom 700 cm^{-1} to 1000 cm^{-1}, adopted from Ref. [16].

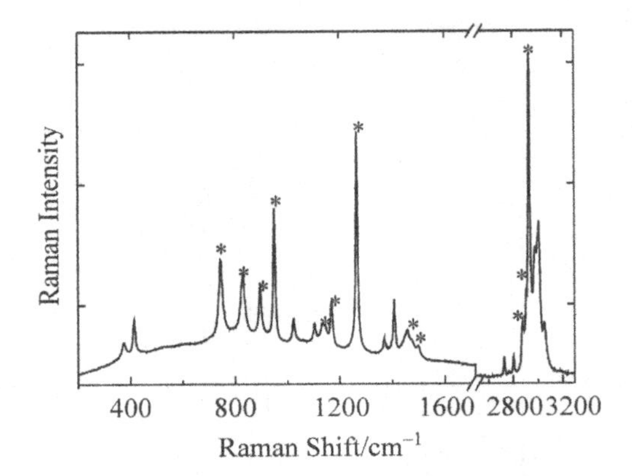

Fig. 6. 8 The Raman spectra of liquid(+)-(R)-methyloxirane by 532 nm excitation. * shows those peak intensities that are employed for the elucidation of the bond polarizabilities.

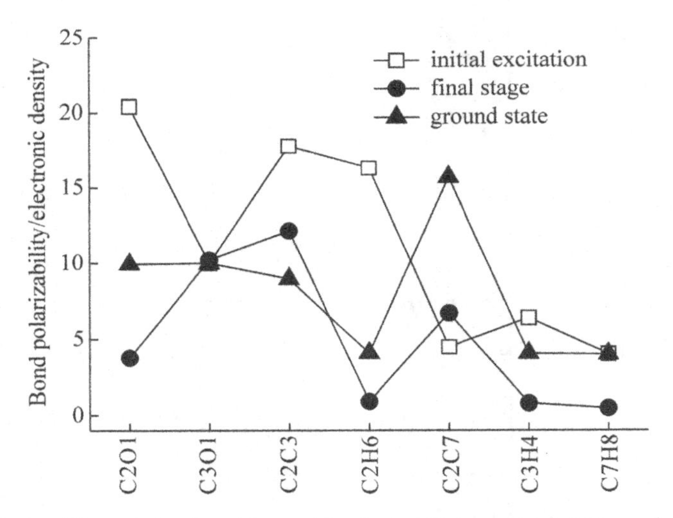

Fig. 6. 9 The relative bond(stretch) polarizabilities of(+)-(R)-methyloxirane at the initial excitation moment(□) and the final stage of Raman relaxation(●) by 532 nm excitation. (▲) is for the calculated bond electronic densities of the ground state. For convenience, the value of C3—O1 is normalized to 10. Note that there is no correlation between the values of the bond polarizability and the calculated bond electronic density.

Table 6. 3 The experimental and fitted Raman shifts in cm^{-1}, relative intensities together with their potential energy distributions of(+)-(R)-methyloxirane under 532 nm. The relative mode intensities are with respect to that at 2932 cm^{-1} which is normalized to 100 for convenience.

Experimetnal	Fitted	Intensity	Potential Energy Distribution
2932	2931	100	S_4(98)
2905	2901	19. 8	S_6(98)
2875	2876	11. 1	S_8(86), S_9(4), S_{10}(11)
1498	1505	2. 3	S_{18}(81), S_3(17)
1407	1410	9. 1	S_3(15), S_5(20), S_{14}(27), S_{23}(31)
1265	1268	38. 8	S_2(3), S_3(25), S_{14}(42)
1168	1171	8. 4	S_5(20), S_{12}(16), S_{13}(19), S_{15}(21)
1133	1130	5. 4	S_1(8), S_2(5), S_{11}(8. 9), S_{16}(71)
949	952	23. 8	S_2(38), S_5(25), S_{19}(26)
895	891	12. 9	S_2(12), S_{11}(11), S_{15}(34), S_{17}(20), S_{20}(26)
828	832	30. 6	S_1(37), S_3(31), S_5(10), S_{15}(5)
745	748	32. 4	S_1(44), S_2(37), S_5(11), S_{19}(6)

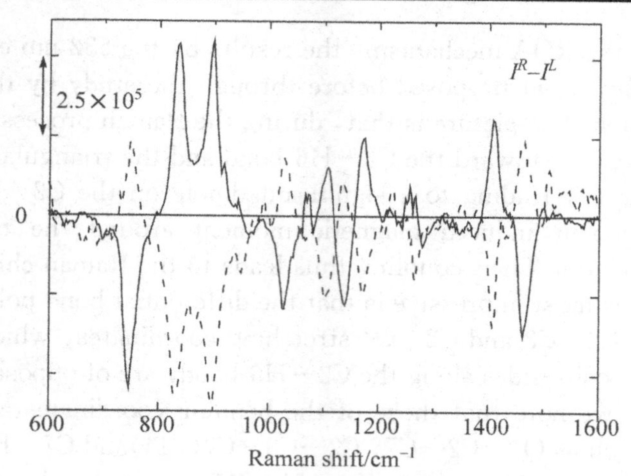

Fig. 6. 10 The ROA spectra of(+)-(R)-methyloxirane by 532. 5 nm excitation. Those of(−)-(S)-methyloxirane are also shown in dashed lines. They are opposite, showing that there are no artifacts.

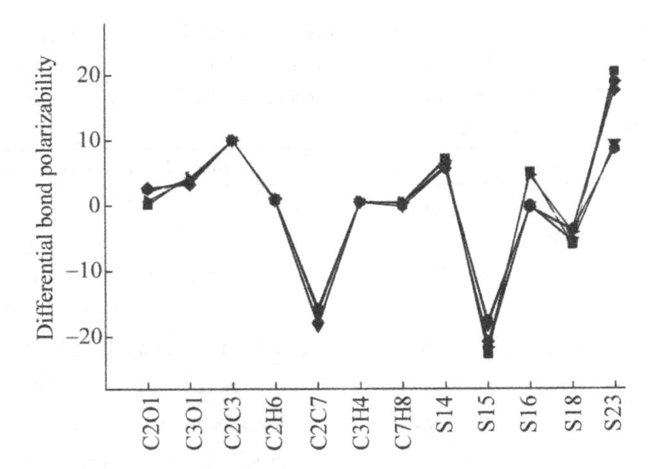

Fig. 6. 11 The relative differential bond polarizabilities of (+)-(R)-methyloxirane under 532. 5 nm excitation by 9 phase set solutions. For convenience, the value of C2—C3 is normalized to 10. S14 involves the O1—C2—C7 and C2—C3—C7 bendings, S15 involves the O1—C3—H_2 and C2—C3—H_2 bendings, S23 involves the C7—H_3 bending, S16 and S18 involve the C3—H_2 wagging(bending).

As to the ROA mechanism, the results by the 532 nm excitation again confirm what proposed before through the study by the 514. 5 nm excitation. The picture is that, during the Raman process, the excited charges are toward the C2—H6 bond and the triangular oxirane skeleton, thus, leading to a significant dipole on the C2—H6 bond and an eminent induced magnetic moment around the triangular oxirane skeleton. Their coupling thus leads to the Raman chirality.

What is most impressive is that the differential bond polarizabilities of the C2—C3 and C2—C7 stretching coordinates, which are on the two opposite sides along the C2—H6 bond, are of opposite signs. Moreover, we note that those of the bending coordinates associated with C7(such as O1—C2—C7, C2—C3—C7(S14)and C7—H_3(S23)) and C3(O1—C3—H_2 and C2—C3—H_2(S15)) respectively are also of opposite signs. This means that the respective responses to the right and left circularly polarized excitations of the portions on the two opposite sides along the C2—H6 bond are in opposite ways. This is probably due to the opposite orientations of the vibrationally induced charge currents along the oxirane ring and the C2—C7—H_3 paths, respectively, in the Raman process. The associated magnetic moments are therefore along opposite orientations. Schematically, this is shown in Fig. 6. 12. We will see that this observation also appears in the alanine case as demonstrated below.

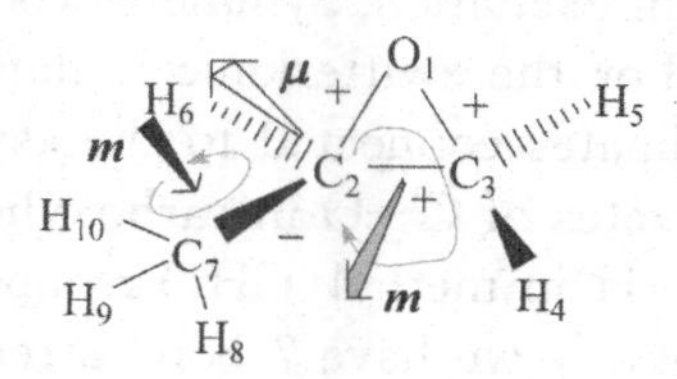

Fig. 6. 12 The signs of differential bond (stretch) polarizabilities for(+)-(R)-methyloxirane and the sketch of the associated vibrationally induced charge currents, the magnetic moments and the induced electric dipoles on the methine C—H bond.

Comments

(1) It is common that there is a C—H bond on the asymmetric C atom in many chiral molecules. The result by methyoxirane shows that in the Raman excitation, charges are highly aggregated on this C—H bond. It will be demonstrated in the following sections that often chiral molecules share this feature. Therefore, this feature, the strong vibrationally induced dipole on this C—H bond, plays a central role in ROA mechanism.

(2) The observation that the differential bond polarizabilities of the stretching coordinates on the two opposite sides along the C2—H6 bond and those of the bending coordinates associated with C7 and C3 respectively are of opposite signs deserves special attention. This will show up again and again in the following chiral systems studied. We will show that, in fact, this belongs to a larger categorical phenomenon called intra-molecular enantiomerism. We will go into its details in Chapter 8.

6.4 The case of L-alanine

Alanine in water solution is zwitterionic. The molecular structure and the atomic numberings of zwitterionic L-alanine are shown in Fig. 6.13. Shown in Fig. 6.14 and Fig. 6.15 are its Raman and ROA spectrum under 532 nm excitation. Symmetry coordinates adopted are shown in Table 6.4. For the zwitterionic L-alanine case, we have 4 bond stretching coordinates connected to the asymmetric center C2. The stretching coordinates of C—O in carboxylic(COO^-), N—H in amino(NH_3^+) and C—H in methyl(CH_3) groups are treated equivalently respectively. Hence, we have 7 bond stretch polarizabilities to elucidate. Besides, we also consider 4 non-negligible intensities whose modes are more of bending coordinates. Correspondingly, we have 11 mode intensities for the analysis. They are shown by * in Fig. 6.14. Table 6.5 shows their Raman shifts, intensities and potential energy distributions

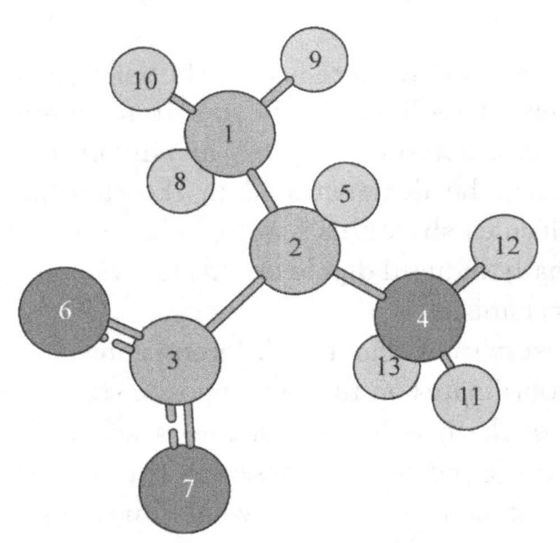

Fig. 6.13 The structure of zwitterionic L-alanine and its atomic numberings. Atoms 6, 7 are oxygen, atom 4 is nitrogen, atoms 1, 2, 3, are carbon and the rest are hydrogen atoms.

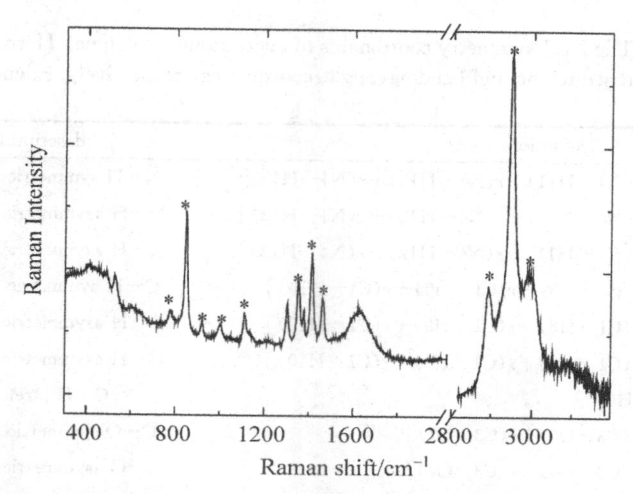

Fig. 6. 14 The Raman spectra of zwitterionic L-alanine in water by 532 nm excitation. * shows those peak intensities that are employed for the elucidation of the bond polarizabilities.

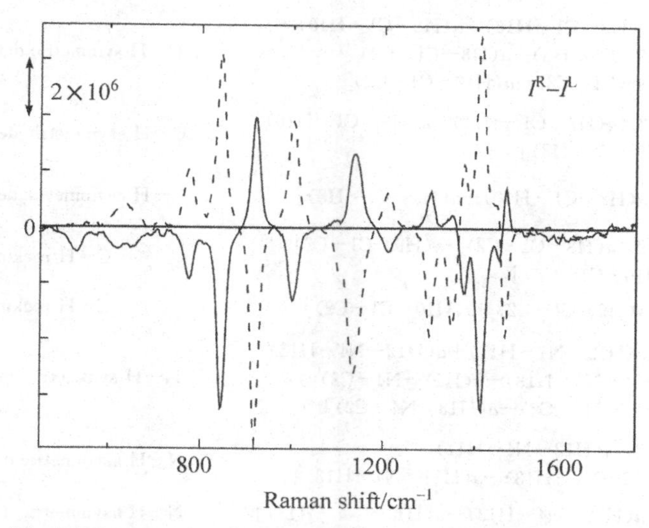

Fig. 6. 15 The ROA spectrum of zwitterionic L-alanine in water by 532. 5 nm excitation. That of D-alanine is also shown in dashed lines. They are opposite, showing that there are no artifacts.

Table 6.4 The local symmetry coordinates of zwitterionic L-alanine. Here r and a denote the bond stretching and bending(angle) coordinates, respectively. τ denotes the dihedral angle.

Definition	description
$S_1=3^{-1/2}[r(\text{N4—H11})+r(\text{N4—H12})+r(\text{N4—H13})]$	N—H symmetric stretching
$S_2=3^{-1/2}[r(\text{N4—H11})-r(\text{N4—H12})+r(\text{N4—H13})]$	N—H asymmetric stretching
$S_3=3^{-1/2}[r(\text{N4—H11})+r(\text{N4—H12})-r(\text{N4—H13})]$	N—H asymmetric stretching
$S_4=3^{-1/2}[r(\text{C1—H8})+r(\text{C1—H9})+r(\text{C1—H10})]$	C—H symmetric stretching
$S_5=3^{-1/2}[r(\text{C1—H8})-r(\text{C1—H9})+r(\text{C1—H10})]$	C—H asymmetric stretching
$S_6=3^{-1/2}[r(\text{C1—H8})+r(\text{C1—H9})-r(\text{C1—H10})]$	C—H asymmetric stretching
$S_7=r(\text{C2—H5})$	C—H stretching
$S_8=2^{-1/2}[r(\text{C3—O6})+r(\text{C3—O7})]$	C—O symmetric stretching
$S_9=2^{-1/2}[r(\text{C3—O6})-r(\text{C3—O7})]$	C—O asymmetric stretching
$S_{10}=r(\text{C1—C2})$	C—C stretching
$S_{11}=r(\text{C2—C3})$	C—C stretching
$S_{12}=r(\text{C2—N4})$	C—N stretching
$S_{13}=6^{-1/2}[a(\text{H9—C1—H10})+a(\text{H8—C1—H10})$ $+a(\text{H9—C1—H8})-a(\text{H8—C1—C2})$ $-a(\text{H9—C1—C2})-a(\text{H10—C1—C2})]$	C—H symmetric deformation
$S_{14}=6^{-1/2}[2\times a(\text{H9—C1—H10})-a(\text{H8—C1—H10})$ $-a(\text{H9—C1—H8})]$	C—H asymmetric deformation
$S_{15}=2^{-1/2}[a(\text{H8—C1—H10})-a(\text{H9—C1—H8})]$	C—H asymmetric deformation
$S_{16}=6^{-1/2}[2\times a(\text{H8—C1—C2})-a(\text{H9—C1—C2})$ $-a(\text{H10—C1—C2})]$	C—H rocking
$S_{17}=2^{-1/2}[a(\text{H9—C1—C2})-a(\text{H10—C1—C2})]$	C—H rocking
$S_{18}=6^{-1/2}[a(\text{H12—N4—H11})+a(\text{H12—N4—H13})$ $+a(\text{H11—N4—H13})-a(\text{H11—N4—C2})$ $-a(\text{H12—N4—C2})-a(\text{H13—N4—C2})]$	N—H symmetric deformation
$S_{19}=6^{-1/2}[2\times a(\text{H12—N4—H11})$ $-a(\text{H12—N4—H13})-a(\text{H11—N4—H13})]$	N—H asymmetric deformation
$S_{20}=2^{-1/2}[a(\text{H12—N4—H13})-a(\text{H11—N4—H13})]$	N—H asymmetric deformation
$S_{21}=6^{-1/2}[2\times a(\text{H11—N4—C2})-a(\text{H12—N4—C2})$ $-a(\text{H13—N4—C2})]$	N—H rocking
$S_{22}=2^{-1/2}[a(\text{H12—N4—C2})-a(\text{H13—N4—C2})]$	N—H rocking

Table 6. 5 The experimental and fitted Raman shifts in cm^{-1}, relative intensities together with their potential energy distributions of zwitterionic L-alanine. The relative mode intensities are with respect to that at 2947 cm^{-1} which is normalized to 100 for convenience. The fit errors of most Raman shifts for both cases are less than 1%.

Experimental	Fitted	Intensity	Potential Energy Distribution
3006	3002	20. 9	S_1 (90), S_3 (9)
2947	2947	100	S_5 (3), S_7 (92)
2890	2889	42. 4	S_4 (60), S_6 (39)
1461	1465	11. 2	S_{13} (10), S_{18} (49), S_{20} (15), S_{26} (13)
1412	1416	22. 7	S_{21} (12), S_{25} (59)
1352	1456	17. 9	S_8 (82), S_9 (17), S_{23} (13)
1111	1103	6. 4	S_{10} (14), S_{17} (29), S_{21} (27)
1000	1005	6. 2	S_{10} (51), S_{17} (14), S_{22} (16)
919	928	2. 5	S_{11} (18), S_{12} (39), S_{16} (20)
845	845	30. 2	S_{11} (19), S_{16} (9), S_{23} (56)
776	767	4. 8	S_{12} (38), S_{16} (49)

The tactics for the phase determination is analogous to the work of methyloxirane. We have only two phase sets remained and they share very parallel bond polarizabilities. For clarity, only one set of the elucidated bond polarizabilities at the initial and final stages(4 ps) of relaxation together with the calculated bond electronic densities of the ground state are shown in Fig. 6. 16 in which the value of C2—C3 bond is normalized to 10. By the phase sets determined and the ROA intensities, the differential bond polarizabilities are readily obtained and are shown in Fig. 6. 17.

The charge excitation is mostly on the methine C2—H5 bond (and C3—O_2) during the Raman process (Fig. 6. 16). It is expected that the coupling of the vibrationally induced dipole on the C2—H5 bond(and even the C3—O_2 bonds) with the vibrationally induced magnetic moments by the charge currents associated with the ring paths through C2—C1—H_3, C3—O_2 and C2—N4—H_3, C3—O_2 plays the ROA role for this system. In this sense, it is not unexpected that the

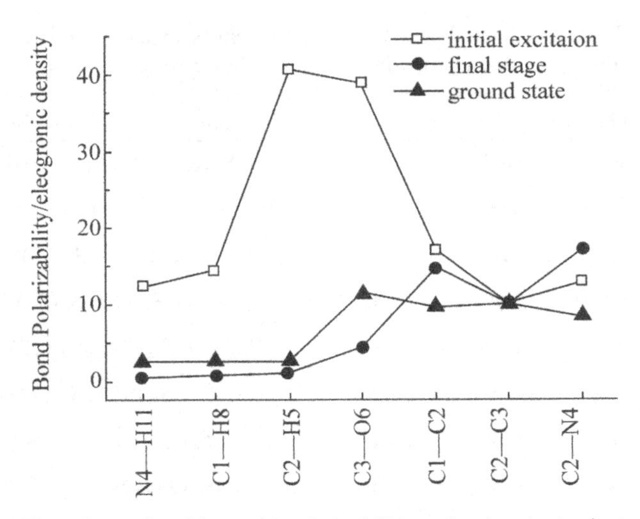

Fig. 6. 16 The relative bond(stretch) polarizabilities of zwitterionic alanine at the initial excitation moment(□) and the final stage(4 ps)(●) of Raman relaxation. (▲) is for the calculated bond electronic densities of the ground state. For convenience, the value of C2—C3 is normalized to 10. Note that there is no correlation between the values of the bond polarizability and the calculated bond electronic density.

asymmetry observed in methyloxirane will also appear in this system. Indeed, as shown in Fig. 6. 17, the differential bond polarizabilities of C1—C2 and C2—N4 stretching coordinates are of opposite signs in one phase set solution and very impartial in the other phase set solution where one is very positively large and the other is very close to zero. Furthermore, we observe that the differential bond polarizabilities due to the bending coordinates associated with C2—C1—H_3 (S16) and those associated with C2—N4—H_3 (S18, S21) are of opposite signs. (That the differential bond polarizabilities of S16 and S17 involving C2—C1—H_3 bending coordinates are of opposite signs shows that the situation could be more complicated than expected as in methyloxirane.) These observations show the asymmetry of the differential bond polarizabilities on the opposite sides along the asymmetric C2—H5 bond. This would not be the case if N4 atom were replaced by the C atom with the loss of chirality. The opposite orientations of the vibrationally induced charge currents along the C2—C1—H_3, C3—O_2

and C2—N4—H_3, C3—O_2 respective paths, which result in the opposite orientations of the associated magnetic moments, is expected to be the physical picture for this observation. Schematically, this is shown in Fig. 6.18.

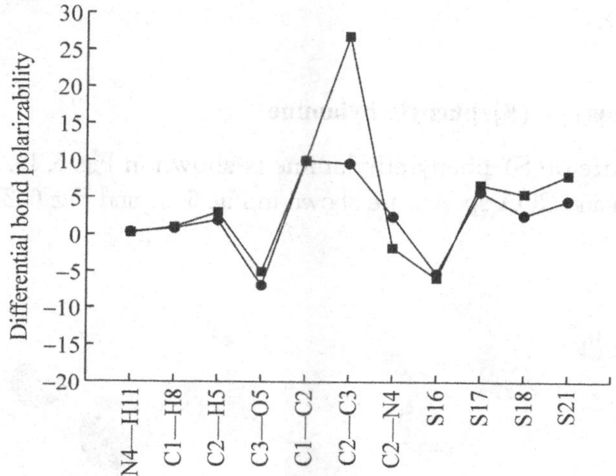

Fig. 6.17 The relative differential bond polarizabilities of zwitterionic L-alanine by 532.5 nm excitation due to various phase set solutions. For convenience, the value of C1—C2 is normalized to 10. S16, S17 involve C2—C1—H_3 bending, S18, S21 involve C2—N4—H_3 bending.

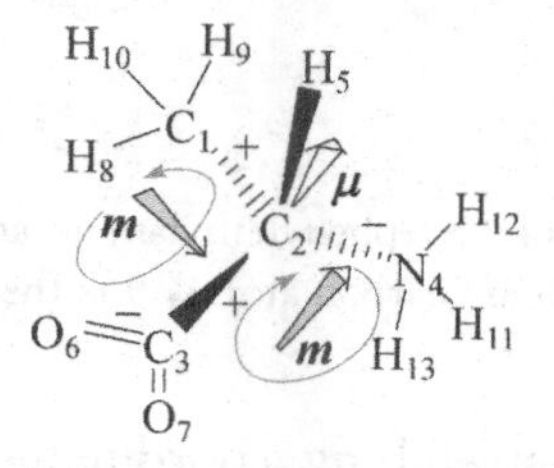

Fig. 6.18 The signs of the differential bond(stretch) polarizabilities for zwitterionic L-alanine and the associated vibrationally induced charge currents. Also shown are the magnetic moments and the induced electric dipoles on the methine C—H bond.

Comments

Through the bond polarizability and differential bond polarizability analysis, we do observe that though methyloxirane and alanine are two molecules with distinctly different structures, they share common ROA features.

6.5 The case of (S)-phenylethylamine

The structure of(S)-phenylethylamine is shown in Fig. 6.19. The measured Raman and ROA spectra are shown in Fig. 6.20 and Fig. 6.21, respectively.

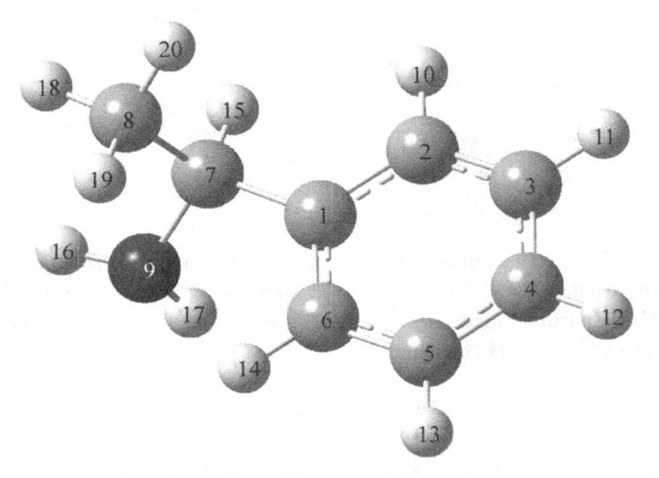

Fig. 6.19 The structure of (S)-phenylethylamine and its atomic numberings. The numberings from 1 to 8 are carbon atoms, 9 is the nitrogen atom and the rest are hydrogen atoms.

We chose 9 bond stretching coordinates (corresponding to the skeletal bonds) and 4 bending coordinates to elucidate their bond polarizabilities. The Raman and ROA intensities associating with these coordinates are generally intense and are shown in Figs. 6.20, 6.21 with * sign except two modes of extremely weak intensity as tabulated in Table 6.6 where the mode wavenumber, Raman, ROA

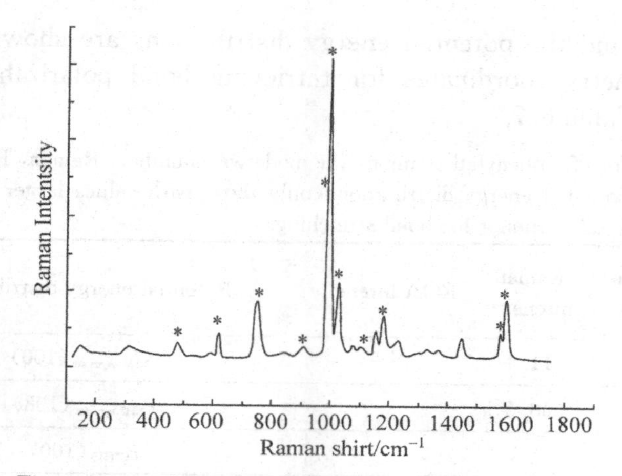

Fig. 6. 20 The Raman spectrum of (S)-phenylethylamine. Those with * are employed for intensity analysis.

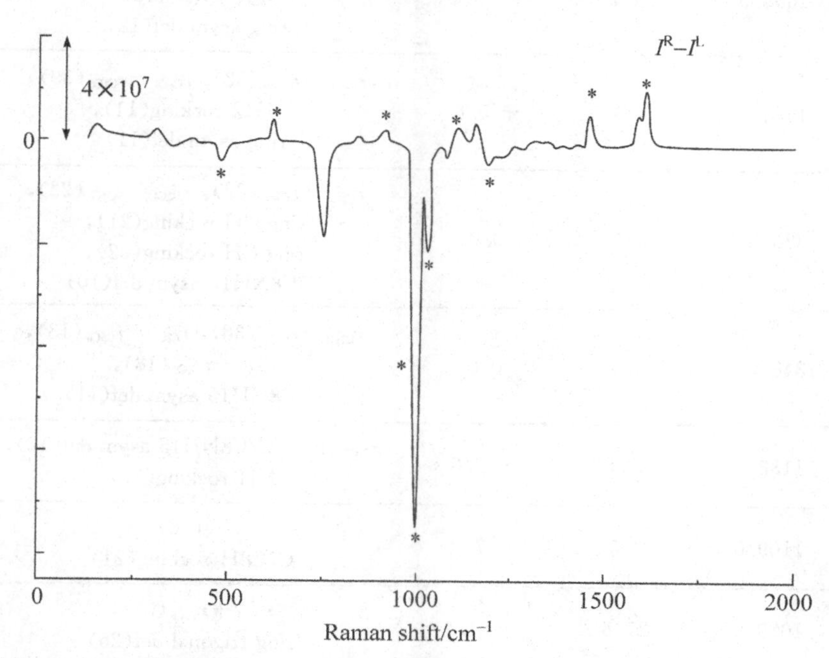

Fig. 6. 21 The ROA spectrum of (S)-phenylethylamine. Those with * are employed for intensity analysis.

intensities and the potential energy distributions are shown. The 13 local symmetry coordinates for retrieving bond polarizabilities are shown in Table 6. 7.

Table 6. 6 For (S)-phenylethylamine, the mode wavenumber, Raman, ROA intensities and the potential energy distributions (only those with values larger than 10 are shown). * by calculation. r for bond stretching.

Experimental wavenumber	Raman intensity	ROA intensity	Potential energy distribution
3372. 0	11. 4	—	$r_{NH2\ asym\ str}$ (100)
3314. 0	63. 5	—	$r_{NH2\ sym\ str}$ (108)
2716. 9	6. 6	—	r_{C7H15} (100)
1604. 5	21. 2	+ 15. 1	$r_{C1C2}+r_{C1C6}$ (18), $r_{C2C3}+r_{C5C6}$ (52), $r_{C3C4}+r_{C4C5}$ (12) ring asym def(12)
1584. 4	6. 9	+ 5. 6	$r_{C1C2}-r_{C1C6}$ (32), $r_{C3C4}-r_{C4C5}$ (39), C4H12 rocking(11), ring asym def(11)
1376. 0 *	0. 0	0. 0	$r_{C2C3}-r_{C5C6}$ (32), $r_{C3C4}-r_{C4C5}$ (22), ring CH rocking(21), ring CH rocking(22), C7C8NH15 asym def(10)
1349. 0 *	0. 0	0. 0	$r_{C1C2}-r_{C1C6}$ (38), $r_{C2C3}-r_{C5C6}$ (13), $r_{C3C4}-r_{C4C5}$ (18), C7C8NH15 asym def(11)
1182. 0	27. 8	− 2. 6	r_{C1C7} (15), C7C8NH15 asym def(12), NH_2 rocking(17)
1109. 0	3. 5	+ 7. 2	r_{C7N} (38), C7C8H_3 rocking(21)
1029. 0	21. 8	− 21. 4	$r_{C3C4}+r_{C4C5}$ (37), ring trigonal def(26)

1001. 0	100. 0	− 100. 0	$r_{C2C3}+r_{C5C6}$ (14), $r_{C3C4}+r_{C4C5}$ (22), ring trigonal def(37),
991. 4	2. 9	0. 0	r_{C7C8} (20), r_{C7N} (23), C7C8H_3 rocking(27), NH_2 rocking(15)
911. 3	5. 0	+ 5. 6	r_{C7C8} (43), r_{C7N} (20), C7C8H_3 rocking(12)
754. 0	48. 3	− 36. 8	$r_{C1C2}+r_{C1C6}$ (15), r_{C1C7} (22), ring asym def(24)
621. 0	7. 1	+ 5. 5	ring asym def(20), C1C7NH15 rocking(11), NH_2 wagging(46)
483. 4	11. 4	− 8. 2	ring asym def(15), C7C8NH15 asym def(36)

Table 6. 7 The symmetry coordinates for (S)-phenylethylamine. *r* stands for bond stretching.

Local symmetry coordinate	In terms of internal coordinates
S_1	$(r_{C1C2}+r_{C1C6})/\sqrt{2}$
S_2	$(r_{C1C2}-r_{C1C6})/\sqrt{2}$
S_3	$(r_{C2C3}+r_{C5C6})/\sqrt{2}$
S_4	$(r_{C2C3}-r_{C5C6})/\sqrt{2}$
S_5	$(r_{C3C4}+r_{C4C5})/\sqrt{2}$
S_6	$(r_{C3C4}-r_{C4C5})/\sqrt{2}$
S_{12}	r_{C1C7}
S_{13}	r_{C7C8}
S_{14}	r_{C7N9}
S_{28}	ring asym def
S_{40}	C7C8NH15 asym def
S_{43}	C1C7NH15 rocking
S_{51}	NH_2 wagging

Considering that the ROA spectrometer can cover only the spectral range below 1800 cm^{-1} which is the region of the stretching and bending motions of the molecular skeleton, and that the coupling of the skeletal motion with C—H and N—H moieties (around 3000 cm^{-1}) is weak, we will pay our attention only to this spectral range for this study. We also note that the couplings of C7—H15, N9—H_2 stretching motions with those of the phenyl C—H and C8—H_3 stretchings (around 3000 cm^{-1}) are very weak. Hence, from the Raman intensities of C7—H15 and N9—H_2 stretching modes (also listed in Table 6. 6), their bond polarizabilities can be readily obtained.

What needs to mention is that the elucidation of the bond polarizabilities of the phenyl C—C stretching coordinates is through the polarizabilities of their symmetric and anti-symmetric coordinates (See Table 6. 7). If the phenyl ring possesses strict symmetries such as D_{6h} group, then the modes with anti-symmetric coordinates will be Raman inactive. In such a case, all the bond polarizabilities of the six phenyl C—C coordinates are identical. In (S)-phenylethylamine, this symmetry is broken, and the Raman peaks due to the anti-symmetric coordinates show up. Just for convenience, this symmetry breaking can be considered as a perturbation. Then the mode intensities due to the symmetric coordinates will possess more diagonal elements of the electronic polarizability tensor while those due to the anti-symmetric coordinates possess more off-diagonal elements of the electronic polarizability tensor. The merit that we will gain as both the symmetric and anti-symmetric coordinates are adopted is that then not only both the differential bond polarizabilities of the symmetric and anti-symmetric coordinates can be obtained but also that their difference demonstrates the differentiability of the diagonal and off-diagonal elements of the electronic polarizability tensor for the chiral asymmetry. This is what we expect. We know in general that in the Raman process, the diagonal elements of the electronic polarizability are larger than the off-diagonal elements. We will see later in our analysis that in ROA , on the contrary, chiral asymmetry is more expressed by the anti-symmetric coordinates or the off-diagonal elements of the electronic polarizability tensor.

The conditions for solving the phases are that all the bond stretching polarizabilities are positive, the deviations among the six phenyl C—C stretching polarizabilities are not two-fold (this value is not so crucial) larger than their average and that the polarizability of the phenyl ring deformation (S28) is less than those of the phenyl C—C stretching polarizabilities. Then, we have seven phase sets left. Among them, three phase sets possess too large deviation (over 10 fold) among the six phenyl C—C stretching polarizabilities. While another phase set is not reasonable since its temporal bond polarizabilities are with too severe variations. Then, only three phase sets remained. The bond polarizabilities corresponding to these three phase sets are shown in Fig. 6. 22. Therein, for comparison, the bond orders by EHMO of the ground state are also shown. Once the phase set is determined, the differential bond polarizabilities can be obtained readily. They are shown in Fig. 6. 23. Quite interestingly, these three phase sets lead to consistent differential bond polarizabilities, thus ensuring the conclusive interpretations for this chiral molecule.

The bond polarizabilities by all the three phase sets demonstrate that in the Raman process, the excited charges are most concentrated on C7—C8 and C1—C7 bonds, especially, on the C7—C8 bond which instead possesses the least charges in the ground state by EHMO calculation. The bond polarizability of N9—H16 (N9—H17) is 0. 69 of those average of the phenyl C—C bonds while EHMO shows this value is 0. 86. Evidently, in the excited virtual state, the charge aggregation on the N9—H16 (N9—H17) bond becomes less. For the C7—H15 bond, its polarizability is 0. 26 of those average of the phenyl C—C bonds while EHMO shows that it is 0. 63. Evidently, in the excited virtual state, the charge aggregation on the C7—H15 bond does not enhance significantly, either.

From Fig. 6. 23, we have the following observations:

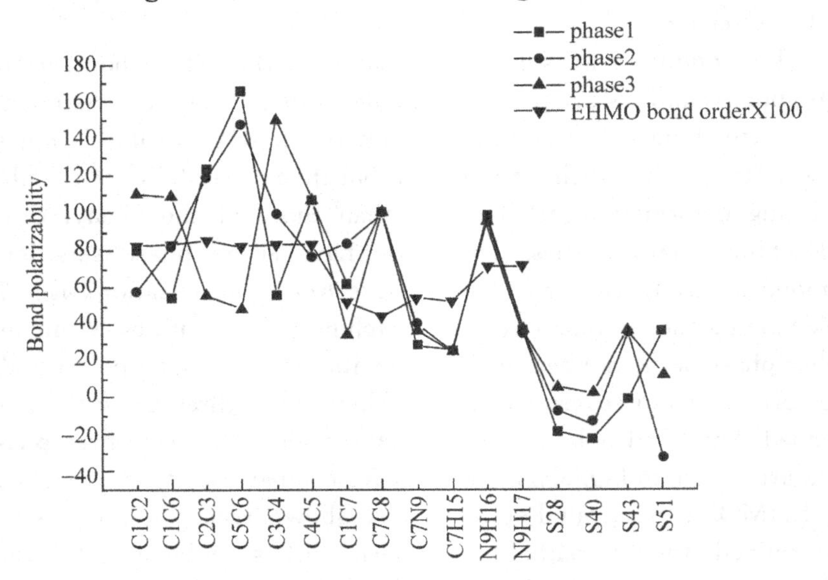

Fig. 6. 22 The bond polarizabilities with those of C7C8 normalized to 100 for convenience. For comparison, EHMO bond orders of the ground state are also shown.

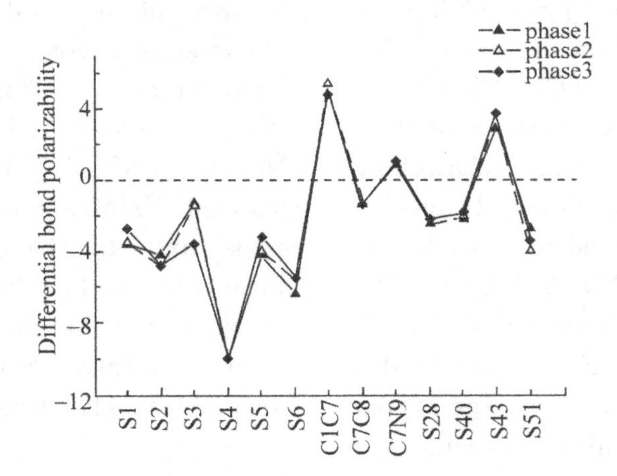

Fig. 6. 23 The differential bond polarizabilities under various phase sets. Those of S4 are normalized to −10 for convenience. Note their consistency.

(1) The anti-symmetric S2, S4, S6 coordinates possess larger differential bond polarizabilities than their respective symmetric coordinates S1, S3 and S5. In other words, the anti-symmetric coordinates possess more ROA or chiral effect. This means that off-diagonal terms of the electronic polarizability tensor show more chiral asymmetry. Specifically, for ROA, we have

$$\Delta\alpha_{xx}, \Delta\alpha_{yy}, \Delta\alpha_{zz} < \Delta\alpha_{xy}, \Delta\alpha_{xz}, \Delta\alpha_{yz}$$

while for Raman process,

$$\alpha_{xx}, \alpha_{yy}, \alpha_{zz} > \alpha_{xy}, \alpha_{xz}, \alpha_{yz}$$

This is not difficult to comprehend. Suppose $\otimes$ is the operation associating with the ROA/chiral process. Then, it is expected that $\otimes$ is commutative for the diagonal terms of the polarizability tensor, say (symbolically), $x \otimes x = x \otimes x$ while for the off-diagonal elements, it can be non-commutative, say, $x \otimes y \neq y \otimes x$, leading to $\Delta\alpha_{xy} \neq 0$. Thus, $\Delta\alpha_{xy} > \Delta\alpha_{xx}, \Delta\alpha_{yy}$. It is expected that the differential bond polarizability of the anti-symmetric coordinate is larger than that of its counterpart, the symmetric coordinate, is a general property.

(2) S4, S6 (both are the phenyl C—C stretchings) and C1—C7 possess larger differential bond polarizabilities, hence more significant ROA effect. As mentioned previously, during the Raman excitation, the excited charges are more concentrated on the C7—C8 bond. Thus the electric dipole by the C7—C8 stretching plays the central role in the ROA electro-magnetic coupling mechanism. The significant differential bond polarizabilities of S4 and S6 show that the magnetic component in the coupling mechanism comes from the phenyl moiety (though we do not have the ROA spectrum for the phenyl C—H, C7—H15 and N—H_2). This can be related to the charge richness in the phenyl ring (the charge flow caused by vibration induces the magnetic moment). We note that the differential bond polarizability of S51 due to NH_2 wagging is larger than those of S28 (phenyl ring asymmetric deformation) and S43 (C1C7NH15 rocking). All these show that around the asymmetric center C7, the chiral effect due to the bending motion is more evident. However, we note that the differential bond polarizabilities of the C7—C8 and C7—N9 stretchings are quite small. This can be attributed to their small magnetic moments since they are

farther away from the phenyl ring though the electric dipole on C7—C8 is significant.

Comments

We notice that Raman process bears more its effect on the symmetric coordinates or the diagonal elements of the electronic polarizability tensor, while ROA bears more its chiral asymmetry on the anti-symmetric coordinates or the off-diagonal elements of the electronic polarizability tensor.

6.6 The case of trans-2, 3-epoxybutane

The structure of trans-2, 3-epoxybutane is shown in Fig. 6. 24. It possesses a C_2 symmetry axis which connects the oxygen atom and the center of the C1—C2 bond. Its Raman and ROA spectra are shown in Fig. 6. 25 and Fig. 6. 26. Table 6. 8 shows the symmetry coordinates employed for the normal mode analysis. The calculation employs Gaussian DFT(b3lyp with 6-31G**). There are 33 normal modes: 17 symmetric(of A representation)and 16 anti-symmetric(of B representation). Table 6. 9 and Table 6. 10 show their Gaussian calculated, observed, fitted mode wavenumbers, Raman and ROA intensities and PED, respectively.

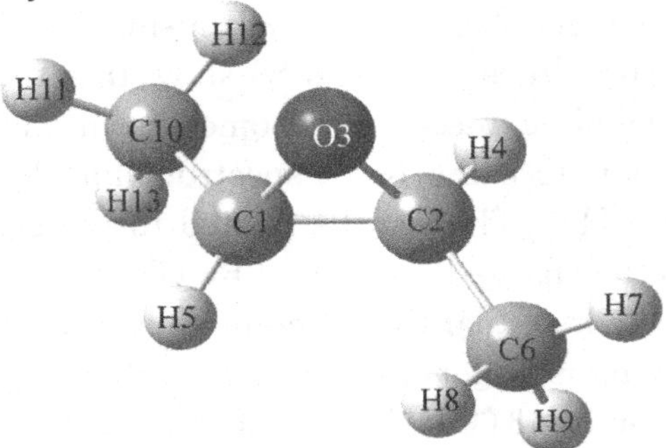

Fig. 6. 24 The structure and atomic numberings of trans-2, 3-epoxybutane. 3 is oxygen, 1, 2, 6, 10 are carbon and the rest are hydrogen atoms. It possesses a C_2 symmetry axis which connects the oxygen atom and the center of the C1—C2 bond.

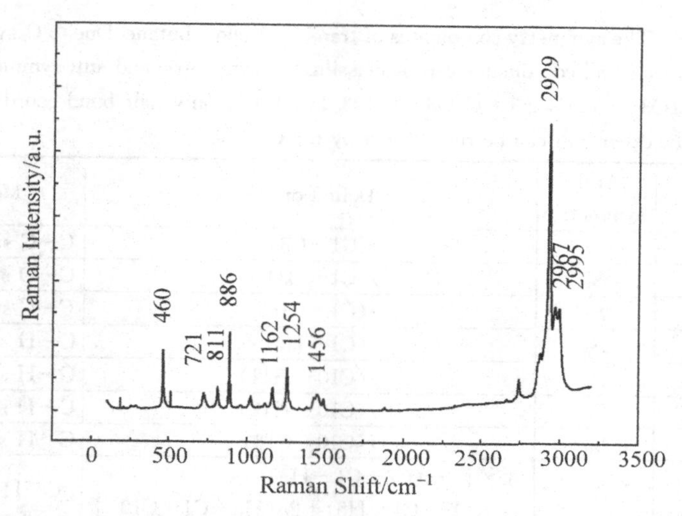

Fig. 6. 25 The Raman spectrum of trans-2, 3-epoxybutane under 532 nm excitation. The labeled peak intensities are those employed for the polarizability analysis.

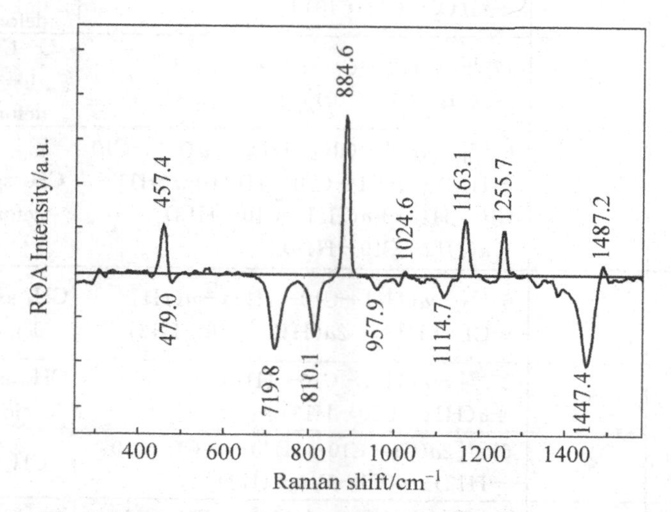

Fig. 6. 26 The ROA spectrum of trans-2, 3-epoxybutane under 532 nm excitation. The labeled peak intensities are those employed for the polarizability analysis.

Table 6.8 The symmetry coordinates of trans-2, 3-epoxybutane. Due to C_2 symmetry, except C1—C2, all coordinates can be classified as symmetric and anti-symmetric such as $S2 = C1O3 + C2O3$, $S3 = C1O3 - C2O3$. For brief, only half bond coordinates are shown. The other half can be related just by the C_2 axis.

Symmetric	Anti-symmetric	Definition	Mode
S_1		$r(C1-C2)$	C—C stretching
S_2	S_3	$r(C1-O3)$	C—O stretching
S_4	S_5	$r(C1-C10)$	C—C stretching
S_6	S_7	$r(C1-H5)$	C—H stretching
S_8	S_9	$r(C10-H11)$	C—H stretching
S_{10}	S_{11}	$r(C10-H12)$	C—H stretching
S_{12}	S_{13}	$r(C10-C13)$	C—H stretching
S_{14}	S_{15}	$6^{-1/2}[-a(C2-C1-H5) -a(O3-C1-H5)+2a(H5-C1-C10)]$	C—H rocking
S_{16}	S_{17}	$2^{-1/2}[a(C2-C1-H5)-a(O3-C1-H5)]$	C—H rocking
S_{18}	S_{19}	$17^{-1/2}[a(C2-C1-C10) +4a(O3-C1-C10)]$	C—C—C and C—C—O deformation
S_{20}	S_{21}	$17^{-1/2}[4a(C2-C1-C10) +a(O3-C1-C10)]$	C—C—C and C—C—O deformation
S_{22}	S_{23}	$6^{-1/2}[-a(C1-C10-H11)-a(C1-C10 -H12)-a(C1-C10-H13)+a(H11-C10-H12)+a(H11-C10-H13) +a(H12-C10-H13)]$	CH_3 symmetric deformation
S_{24}	S_{25}	$6^{-1/2}[-a(H11-C10-H12)-a(H11 -C10-H13)+2a(H12-C10-H13)]$	CH_3 asymmetric deformation
S_{26}	S_{27}	$2^{-1/2}[-a(H11-C10-H12) +a(H11-C10-H13)]$	CH_3 asymmetric deformation
S_{28}	S_{29}	$6^{-1/2}[2a(C1-C10-H11)-a(C1-C10 -H12)-a(C1-C10-H13)]$	CH_2 rocking
S_{30}	S_{31}	$2^{-1/2}[a(C1-C10-H12)-a(C1-C10 -H13)]$	CH_2 rocking
S_{32}	S_{33}	$\tau(O3-C2-C6-H7)$	CH out of OCC plane bending

Table 6.9 The Gaussian calculated, observed, fitted mode wavenumbers, Raman and ROA intensities and PED of the symmetric modes of trans-2, 3-epoxybutane, Raman intensity at 2981(cm^{-1}) is set to 100, ROA intensity at 1444(cm^{-1}) is set to 100.

Gaussian /cm^{-1}	Observed /cm^{-1}	Fitted /cm^{-1}	Raman intensity	ROA intensity	PED
3009	2995	2997	40	—	S_{10}(70)S_{12}(12)S_8(11)S_6(7)
2989	2967	2967	44	—	S_{12}(53)S_8(47)
2981	2929	2931	100	—	S_6(93)S_{10}(7)
2927	2901	2902	22	—	S_8(43)S_{12}(34)S_{10}(23)
1479	1487	1485	3	18	S_{14}(24)S_1(22)S_{26}(21) S_{30}(10)S_4(9)S_2(8)
1452	1456	1456	5	—	S_{24}(89)S_{28}(6)
1434	1425	1425	4	—	S_{26}(69)S_{14}(13)S_1(6)S_4(5)
1378	1380	1380	1	−4	S_{22}(93)S_4(2)
1254	1254	1254	11	28	S_{14}(55)S_1(23)S_2(9)S_{20}(5)
1162	1162	1162	7	52	S_{16}(45)S_{28}(18)S_{18}(11) S_2(7)S_4(7)S_{30}(5)
1115	1112	1113	4	−16	S_{30}(44)S_{20}(18)S_4(15)S_2(9)S_{16}(7)
1019	1021	1021	5	4	S_{16}(42)S_{28}(37)S_4(5)S_{18}(5)S_2(5)
892	886	888	12	80	S_2(44)S_{28}(26)S_4(22)
807	811	810	7	−38	S_1(34)S_{30}(28)S_4(17)S_2(13)
462	460	460	10	41	S_{20}(63)S_4(15)S_{18}(8)S_1(6)S_{30}(5)
262	258	259	1	−48	S_{18}(78)S_{32}(9)S_{28}(5)
206	—	—	—	—	S_{32}(101)

Table 6. 10 The Gaussian calculated, observed, fitted mode wavenumbers, Raman and ROA intensities and PED of the anti-symmetric modes of trans-2, 3-epoxybutane, Raman intensity at 2981 (cm^{-1}) is set to 100, ROA intensity at 1444 (cm^{-1}) is set to 100.

Gaussian (cm^{-1})	Observed (cm^{-1})	Fitted (cm^{-1})	Raman intensity	ROA intensity	PED
3012	—	—	—	—	S_{11} (63) S_7 (17) S_{13} (12)
2988	—	—	—	—	S_{13} (54) S_9 (47)
2985	—	—	—	—	S_7 (83) S_{11} (14)
2926	—	—	—	—	S_9 (43) S_{13} (34) S_{11} (23)
1457	1456	1456	5	—	S_{25} (71) S_{27} (17)
1444	1443	1443	4	−100	S_{27} (73) S_{25} (19)
1382	1380	1380	1	—	S_{23} (91) S_5 (12)
1331	1335	1334	1	−10	S_{15} (84)
1148	—	—	—	—	S_{17} (63) S_{31} (18)
1100	—	—	—	—	S_5 (40) S_{29} (32) S_{19} (14)
1012	—	—	—	−3	S_5 (40) S_{29} (32) S_{19} (14)
951	958	967	2	−14	S_{31} (61) S_{17} (32) S_{15} (10)
741	722	728	8	−56	S_3 (90)
475	—	—	—	−7	S_{19} (79) S_{29} (15)
294	286	289	2	−5	S_{21} (76) S_9 (13)
224	—	—	—	—	S_{33} (94)

A. The symmetric coordinates

The criteria for the phase determination are that the bond polarizabilities of C—C and C—H are positive and in the temporal domain, they decay monotonically. Thus, we have 10 phase sets left. Fig. 6. 27 and Fig. 6. 28 are the (symmetric) bond polarizabilities and the (symmetric) differential bond polarizabilities.

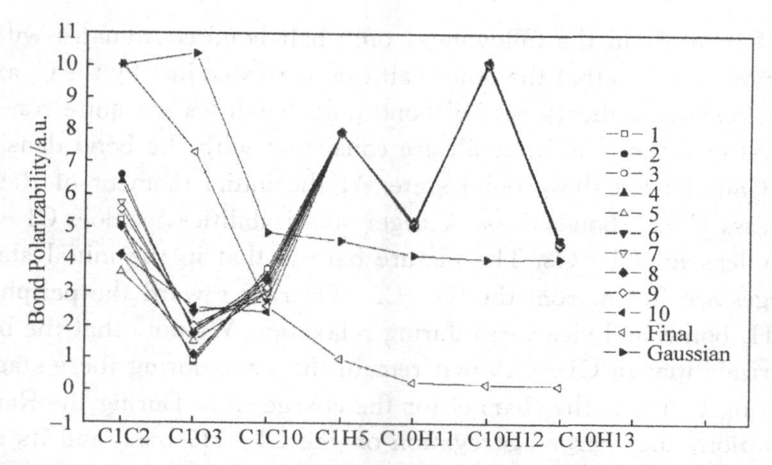

Fig. 6.27 The symmetric bond polarizabilities of trans-2, 3-epoxybutane for the 10 phase sets at the initial time, after relaxation (Final) and the bond densities of the ground state by Gaussian software.

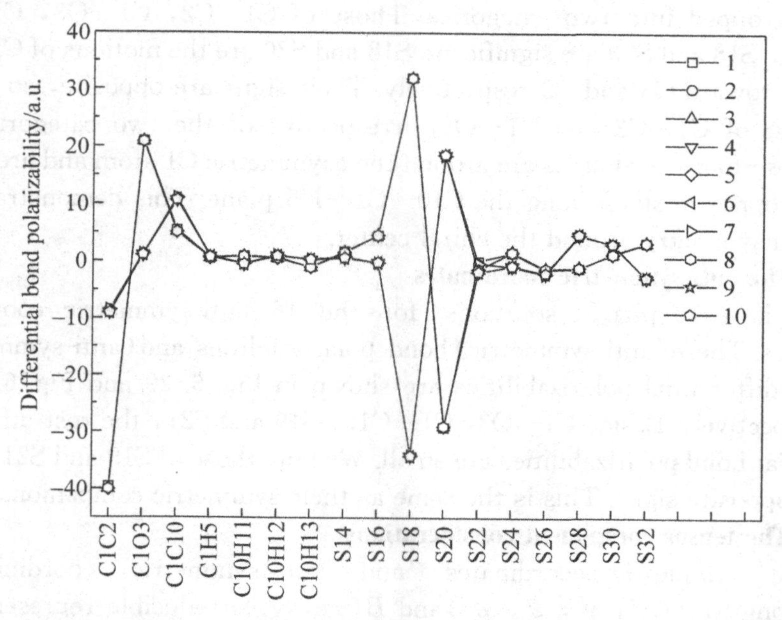

Fig. 6.28 The symmetric differential bond polarizabilities of trans-2, 3-epoxybutane.

For brief, in the followings, only half bond coordinates will be mentioned due to that the other half can be related just by the C_2 axis.

As shown, the 10 sets of bond polarizabilities are quite consistent. After relaxation, they all are consistent with the bond densities (by Gaussian) in the ground state. At the initial moment of Raman process, C—H_3 bonds possess larger polarizabilities, so does C1—H5 while less for C1—O3. The picture here is that in the initial stage, charges are flown from the C1—C2—O3 ring toward the peripheral C—H_3 bonds and vice versa during relaxation. We note that the bond polarizabilities of C1—C10 just remain the same during these stages, showing that it is the channel for the charge flow. During the Raman excitation, the charge aggregation on the C1—H5 bond and its coupling with the magnetic dipole by the C1—C2—O3 ring vibration can play the central role of the ROA mechanism.

The differential bond polarizabilities due to different phases can be grouped into two categories. Those of C1—C2, C1—O3, C1—C10, S18 and S20 are significant. S18 and S20 are the motions of C1—C10 toward O3 and C2 respectively. Their signs are opposite, so are those of C1—C2 and C1—O3, irrespective of the two categories. These four coordinates are around the asymmetric C1 atom and are on the opposite sides along the C10—C1—H5 plane. This demonstrates the asymmetry around the chiral center.

B. The anti-symmetric coordinates

We have 13 phase solutions for the 16 anti-symmetric coordinates. These (anti-symmetric) bond polarizabilities and (anti-symmetric) differential polarizabilities are shown in Fig. 6. 29 and Fig. 6. 30 respectively. Beside C1—O3, C1—C10, S19 and S21, the rest differential bond polarizabilities are small. We note those of S19 and S21 are of opposite signs. This is the same as their symmetric companions.

C. The tensor component considerations

The symmetric coordinates and anti-symmetric coordinates belong to $A(x^2, y^2, z^2, xy)$ and $B(xz, yz)$ irreducible representations, respectively. (The z direction is defined as along O3 and the mid-point of C1—C2.)

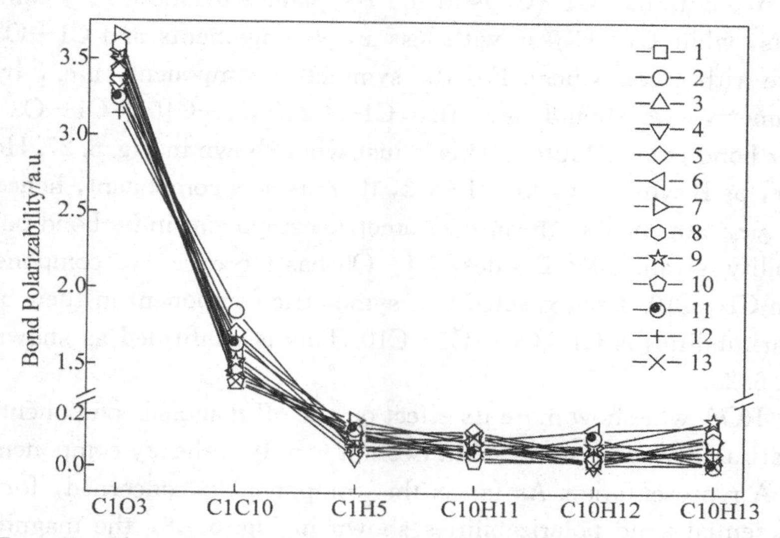

Fig. 6. 29 The anti-symmetric bond polarizabilities of trans-2, 3-epoxybutane.

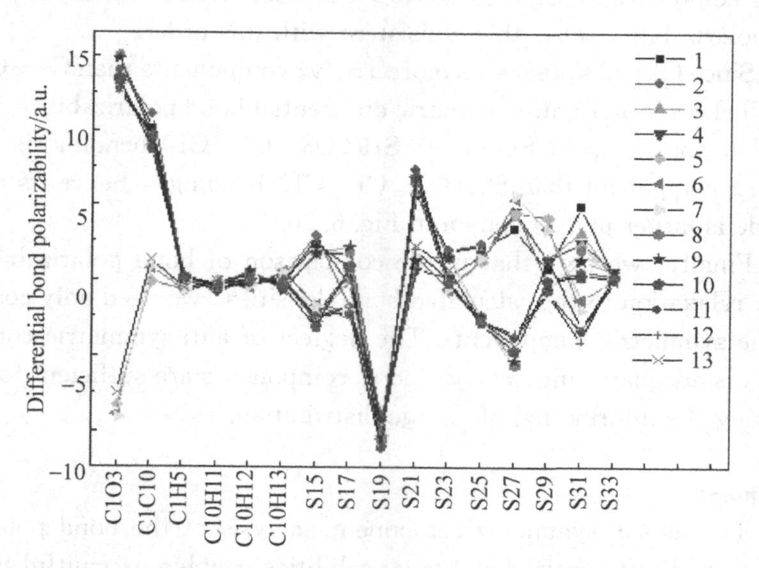

Fig. 6. 30 The anti-symmetric differential bond polarizabilities of trans-2, 3-epoxybutane.

We note that C1—C2 is in the x-y plane with more x, y components, while C1—C10 is with less x, y components and C1—O3 is more with z component. For the symmetric component, i. e., by A symmetry, we should have that C1—C2 > C1—C10 > C1—O3 for their bond polarizabilities. This is just what shown in Fig. 6. 27. However, by B symmetry, for C1—C2, there is no z component, hence no xz, yz components, the anti-symmetric component in its bond polarizability is vanishing. Besides, C1—O3 has more xz, yz components than C1—C10. The expected anti-symmetric component in their bond polarizabilities is C1—O3>C1—C10. This is confirmed as shown in Fig. 6. 29.

ROA will show more its effect on the off-diagonal components of polarizability as was noticed in Section 6. 5. It is the xy component in the A representation. As far as this component is concerned, for the differential bond polarizabilities shown in Fig. 6. 28, the magnitude order could be C1—C2>C1—C10>C1—O3. Hence, the most possible polarizability set is that consistent with this order.

Since C1—O3 possesses more xz, yz components than C1—C10, it will show larger anti-symmetric differential bond polarizability. This is what shown up in Fig. 6. 30. S19 (O3—C1—C10 bending) is with more z component than S21(C2—C1—C10 bending), hence its magnitude is larger just as shown in Fig. 6. 30.

Finally, we note that in the comparison of bond polarizabilities after relaxation to the calculated bond densities, we need only consider the symmetric components. The neglect of anti-symmetric components is adequate since x^2, y^2 and z^2 components are sufficient for retrieving the information of charge distribution.

Comments

The tensor/symmetry component analysis of the bond polarizabilities and differential bond polarizabilities enables us fruitful information concerning chirality that is embedded in the Raman and ROA intensities.

References

[1] Mislow K. Top. Stereochem. , 1999, 22:1.

[2] Lough W J, Wainer I W. Chirality in Natural and Applied Science. Oxford: Blackwell, 2002.

[3] Eliel E L, Wilen S H, Doyle M P. Basic Organic Stereochemistry. New York: John Wiley and Sons, 2001.

[4] Buckingham A D. Faraday Discuss. , 1994, 99: 1.

[5] Nafie L A, Che D. Adv. Chem. Phys. , 1994, 85: 105.

[6] Barron L D, Hecht L, Bell A F, et al. Appl. Spectrosc. , 1996, 50: 619.

[7] Barron L D, Hecht L, Bell A D. Vibrational Raman Optical Activity of Biomolecules. In Circular Dichroism and the Conformational analysis of Biomolecules, G. D. Fasman(Ed), New York: Plenum, 1996.

[8] Blanch E W, Morozova-Roche L A, Cochran D A E, et al. J. Mol. Biol. , 2000, 301: 553.

[9] Josef K, Vladimr B, Vladimr K, et al. J. Phys. Chem. , A 2006, 110: 4689.

[10] Barron L D, Gargaro A R, Hecht L, et al. Spectrochim. Acta A, 1991, 47A: 1001.

[11] Polavarapu P L, Hecht L, Barron L D. J. Phys. Chem. , 1993, 97:1793.

[12] Polavarapu P L, Bose P K, Hecht L, et al. J. Phys. Chem. , 1993, 97: 11211.

[13] Polavarapu P L, Black T M, Barron L D, et al. J. Am. Chem. Soc. , 1993, 155: 7736.

[14] Freedman T B, Balukjian G A, Nafie L A. J. Am. Chem. Soc. , 1985, 107: 6213.

[15] Freedman T B, Lee E, Nafie L A. J. Phys. Chem. A, 2000, 104: 3944.

[16] Bose P K, Polavarapu P L, Barron L D, et al. J. Phys. Chem. , 1990, 94: 1734.

Chapter 7

More applications on ROA

7.1 The case of (S)-2-amino-1-propanol

The structure of (S)-2-amino-1-propanol was optimized by Gaussian09W on the level of DFT B3LYP with basis set aug-ccpVDZ. Shown in Fig. 7.1 is its structure and atomic numberings. Its vibrational mode wavenumbers, Raman and ROA intensities together with PED (potential energy distribution), $\beta(G')^2$ and $\beta(A)^2$ parameters which respectively describe the magnetic and quadrupole effects of ROA under 532 nm excitation were also calculated. These parameters for the elucidation of bond polarizabilities and differential bond polarizabilities are listed in Table 7.1. The symmetry coordinates for the normal mode analysis are listed in Table 7.2.

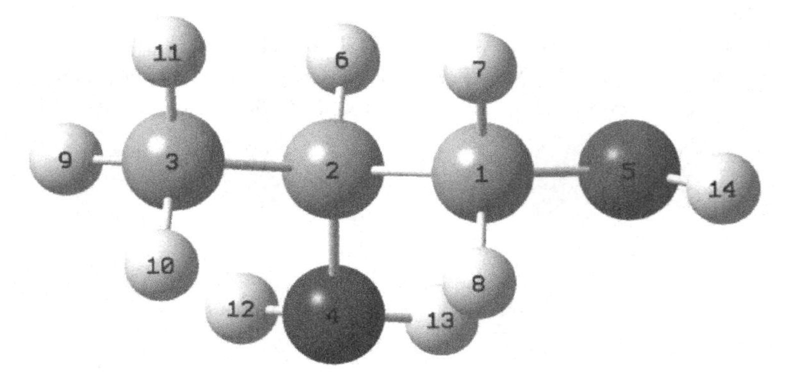

Fig. 7.1 The structure of (S)-2-amino-1-propanol and its atomic numberings. Atoms labeled as 1, 2, 3 are carbon, 4 is nitrogen, 5 is oxygen and the rest are hydrogen. Note that H9—C3—C2—C1—O5—H14 lie more or less on a plane and C2—H6 bond protrudes out of the plane.

Table 7.1 The Raman wavenumbers, Raman and ROA intensities(under 532 nm excitation, ICPu/SCPu(180°)configuration), PED(only those larger than 10 are shown) and ROA parameters, $\beta(G')^2$, $\beta(A)^2$ (10^4 Å^5/amu) of the vibrational modes employed for the elucidation of bond polarizabilities and differential polarizabilities.

Wavenumber	Raman intensity	ROA intensity	PED	$\beta(G')^2$	$\beta(A)^2$
3837.2	782.2	89.2	S13(100)	0.79	0.39
3563.4	557.1	708.2	S12(100)	7.28	0.26
3491.3	1222.5	−85.8	S11(127) S32(15)	−0.98	0.26
3093.1	709.7	−35.8	S10(77) S9(22)	0.21	−1.77
3087.0	609.1	−255.5	S9(78) S10(21)	−2.78	0.37
3043.5	588.3	1022.5	S7(98)	7.29	10.06
3023.6	1834.3	−109.5	S8(98)	−0.98	−0.45
3000.7	919.5	−93.2	S6(99)	−0.14	−2.47
2947.4	2177.4	−795.9	S5(99)	−4.12	−12.5
1143.6	34.0	45.8	S3(36) S16(23) S28(17)	0.39	0.25
1038.7	72.8	135.5	S1(16) S4(72) S30(11)	1.20	0.61
1026.4	13.8	−83.0	S2(26) S3(10) S4(20) S27(14) S30(17)	−0.76	−0.30

953.8	12.3	−90.3	S1(15) S2(14) S16(23) S28(28)	−0.77	−0.49
943.1	28.5	−173.9	S3(10) S4(11) S16(14) S27(25) S32(10)	−1.54	−0.80
877.3	83.0	−77.8	S3(16) S32(101)	−0.67	−0.40
831.9	45.7	150.3	S1(21) S2(10) S3(41) S32(10)	1.15	1.24

Table 7.2 The local symmetry coordinates in terms of the internal coordinates. r denotes the bond stretching coordinate.

Symmetry coordinate	Expression in internal coordinates
S1	r_{C1C2}
S2	r_{C2C3}
S3	r_{C2N4}
S4	r_{C1O5}
S5	r_{C2H6}
S6	$(r_{C1H7}+r_{C1H8})/\sqrt{2}$
S7	$(r_{C1H7}-r_{C1H8})/\sqrt{2}$
S8	$(r_{C3H9}+r_{C3H10}+r_{C3H11})/\sqrt{3}$
S9	$(2r_{C3H9}-r_{C3H10}-r_{C3H11})/\sqrt{6}$
S10	$(r_{C3H10}-r_{C3H11})/\sqrt{2}$

S11	$(r_{N4H12}+r_{N4H13})/\sqrt{2}$
S12	$(r_{N4H12}-r_{N4H13})/\sqrt{2}$
S13	r_{O5H14}
S16	$C1H_2$ rocking
S27/S28	$C3H_3$ rocking
S32	C1 out of NH_2 plane wagging

It is found that the normal modes due to C—H, N—H and O—H bond stretchings are decoupled from the skeletal C—C, C—N and C—O bond stretching and bending motions. Hence, for the elucidation of bond polarizabilities, we will treat these two motions separately. For the former case, we have 9 bond(stretch) polarizabilities(corresponding to S5, S6, S7, S8, S9, S10, S11, S12, S13) to determine while for the skeletal motion, we have 4 bond(stretch) polarizabilities (corresponding to S1, S2, S3, S4) to determine. Since these 4 bond stretchings are coupled to the C1—H_2 rocking (S16), C3—H_3 rocking(S27, S28) and C2 out of N4—H_2 wagging (S32), for this case, totally we have 7 bond(stretch and bend) polarizabilities to determine. Here, we have either S27 or S28 to choose for C3—H_3 rocking since they are rather equivalent. We have tried both choices and the results are consistent. For brief, only the choice of S28 is presented hereby.

The criterion for the phase choice is that the bond stretching polarizability of the individual bond coordinate(C—H, N—H, O—H, C—C, C—N, C—O bonds) is positive. (Those of the C1—H, C3—H, N4—H coordinates can be deduced from those of the symmetry coordinates, like S6 to S12.) For the case of C—H, N—H and O—H bond stretchings, there are 4 phase sets. For the skeletal case, there are 13 phase sets that satisfy this condition. The bond polarizabilities and differential bond polarizabilities of these coordinates are shown in Table 7. 3(a) and Table 7. 3(b) respectively.

From Table 7.3(a), we conclude immediately the following observations, regardless of the un-determined phase sets:

(1) The polarizability of the symmetric coordinate is larger than that of its counter anti-symmetric coordinate, in magnitude(by notation, S6>S7; S8>S9, S10; S11>S12). While for the differential polarizability, we have that of the symmetric coordinate is less than its counter anti-symmetric one (S6<S7; S8<S9, S10; S11<S12.). Furthermore, the differential polarizabilities of the symmetric coordinates(S6, S8, S11)are negative. This shows that though the symmetric mode bears more Raman effect, it is the anti-symmetric mode that bears more ROA implication.

(2) The differential bond polarizability of C2—H6(S5) is negative, while that of O5—H14(S13) is positive, independent of the undetermined phase sets. Their stereo-structural significance will be clear in the following analysis.

Table 7.3(a) The relative bond polarizabilities(upper row) and differential bond polarizabilities (lower row) for the various phase sets.

Phase set label	S5	S6	S7	S8	S9	S10	S11	S12	S13
1	98	81	39	84	−78	−9	84	59	75
	−2.6	−0.4	9.2	−0.1	2.1	−0.3	−0.7	7.4	0.8
2	89	76	−62	83	−78	−17	84	59	75
	−4.1	−1.2	−8.4	−0.4	2.2	−1.7	−0.7	7.4	0.8
3	98	81	39	84	−78	−9	86	−56	75
	−2.6	−0.4	9.2	−0.1	2.1	−0.3	−0.4	−7.4	0.8
4	89	76	−62	83	−78	−17	86	−56	75
	−4.1	−1.2	−8.4	−0.4	2.2	−1.6	−0.4	−7.4	0.8

Table 7. 3 (b) The relative bond polarizabilities (upper row) and differential bond polarizabilities (lower row) for the various phase sets.

Phase set label	S1	S2	S3	S4	S16	S28	S32
1	19.1	37.7	5.5	20.5	9.4	3.9	−34.8
	3.0	1.5	0.2	8.8	−2.1	−4.6	7.9
2	6.8	42.0	10.0	16.6	5.5	−1.8	−41.7
	11.9	−1.5	−3.0	11.7	0.6	−0.4	12.9
3	15.7	15.7	18.5	25.8	10.6	−1.2	−54.4
	5.0	14.8	−7.6	5.7	−2.8	−1.5	19.6
4	3.4	20.0	23.0	21.9	6.7	−7.0	−6.1
	14.0	11.7	−10.8	8.6	−0.1	2.6	24.7
5	21.0	41.6	20.3	21.8	−0.9	13.5	−26.7
	3.2	2.1	2.2	9.0	−3.5	−3.3	9.0
6	1.3	32.4	1.1	21.5	−4.7	3.0	66.6
	2.7	−2.1	14.0	4.2	4.2	1.4	−16.0
7	8.7	45.9	24.8	17.8	−4.7	7.7	−33.6
	12.2	−1.0	−1.0	11.9	−0.7	0.8	14.0
8	37.3	24.2	8.5	41.2	−9.3	8.0	−6.1
	−6.6	12.5	9.5	−2.7	1.6	−0.1	−3.6
9	10.2	6.1	9.7	30.7	0.2	3.7	54.0
	−4.1	14.2	9.4	−1.7	0.7	0.3	−9.2
10	17.6	19.6	33.4	27.0	0.2	8.3	−46.2
	5.3	15.3	−5.6	5.8	−4.2	−0.2	20.7
11	25.0	28.5	12.9	37.3	−13.1	2.2	−13.0
	2.2	9.3	6.2	0.1	4.4	4.1	1.4
12	5.3	23.8	37.8	23.1	−3.5	2.5	−53.1
	14.2	12.2	−8.8	8.7	−1.4	3.9	25.8
13	52.7	1.8	7.9	2.0	−13.1	3.2	−8.5
	−3.8	8.3	9.4	−10.0	0.8	−0.9	−4.0

The vibrational modes by these C—H, N—H and O—H bond stretchings are, in fact, very localized as shown by their PED's in Table 7. 1. Hence, for these modes, the signs of their differential bond polarizabilities should be consistent with those of their $\beta(G')^2 + \beta(A)^2/3$[1,2] (This is a quantum mechanical quantity which determines the ROA mode intensity sign under backward scattering configuration). This offers us one more criterion for the nail-down of the phase set. Then, the first phase set in Table 7.3(a) survives (except the sign of the differential polarizability for S9) better than the rest phase sets.

The 13 un-determined phase sets for the skeletal bond coordinates seem to be a cumbersome task to resolve. However, careful scrutiny over these results [Table7.3(b)] leads to the conclusion that the signs of the differential bond polarizabilities of C1—C2 (S1) and C1—O5 (S4) are the same, regardless of the phase sets. Furthermore, the mode at 1038.7 cm^{-1} is mainly of C1-C2 and C1-O5 stretches (their PED sum is 88) and its $\beta(G')^2 + \beta(A)^2/3$ is positive. This leads to the inference that the signs of differential bond polarizabilities of C1—C2 and C1—O5 are also positive. In fact, this is consistent with the observation that in the 13 un-determined phase sets, 10 phase sets are with those of C1—C2 and C1—O5 positive, i.e., it is highly possible that the proper phase set would lead to positive differential bond polarizabilities for C1—C2 and C1—O5.

By far, we have nailed down the differential polarizability signs for C1—C2 (S1), C1—O5 (S4) and O5—H14 (S13) (these are positive) and C2—H6 (S5) (this is negative) coordinates. It is very interesting to note that C1—C2, C1—O5 and O5—H14 bonds, more or less, lie on a common plane (see Fig. 7.1) and that they share the same differential polarizability sign, while C2—H6, which protrudes out of this plane, possesses the opposite differential bond polarizability sign. Since C2—C3 also lies on this plane, we therefore presume its differential bond polarizability sign (S2) could be also positive. In fact, this is consistent with the observation that among the 13 un-determined phase sets, 10 are with that of C2—C3 positive.

Another point that deserves mention is that, as shown in Table 7.3(a) and (b), the bond polarizabilities of the C—H, N—H and O—H bonds are significantly larger than those of the skeletal (C—C, C—N C—O) bonds. This is an indication that during Raman excitation, the disturbed/excited charges just tend to spread to the molecular periphery due to electronic repulsion. Notably, the bond polarizability of C2—H6 (Table 7.3(a), S5) is the largest. The highly concentrated charges on the asymmetric C—H bond and consequently its large vibrational dipole are believed to play the major role in the ROA mechanism.

Comments

(1) As far as the polarizability is concerned, the symmetric and anti-symmetric coordinates are respectively more significant in the Raman and ROA processes. This was also observed in the case of (S)-phenylethylamine (Section 6.5).

(2) For (S)-2-amino-1-propanol, those bonds lying on a common plane share the same differential bond polarizability sign while that of the asymmetric C—H bond, which protrudes out of the plane, possesses the opposite sign. This demonstrates that ROA can offer more stereo-structural implication than Raman. Therefore, differential bond polarizability, which is relating to ROA intensity, is the appropriate parameter for interpreting the three dimensional configuration of a molecule.

(3) Though user-friendly, Gaussian09W is more or less like a 'black box' that why and how it does so are often not very clear to most users. The physics behind its outputs may not be so evident. The bond polarizability analysis is indeed an appropriate way to comprehend the physical meanings and picture of its outputs.

7.2 The case of (S)-1-amino-2-propanol

The configuration of (S)-1-amino-2-propanol was optimized by Gaussian09W at the DFT/B3LYP/aug-cc-pVDZ level. Its optimized structure is shown in Fig. 7. 2(a). The symmetry coordinates are defined in Table 7. 4. Table 7. 5 shows the Raman wavenumbers, Raman and ROA intensities(under 532 nm excitation, ICPu/SCPu(180°) configuration), PED and ROA parameters, $\beta(G')^2$, $\beta(A)^2/3$. $\beta(G')^2$ is the parameter describing the coupling mechanism of the electric and magnetic dipoles while $\beta(A)^2/3$ is that of the electric dipole and quadrupole. For this backward scattering, the ROA intensity is just proportional to $\beta(G')^2+\beta(A)^2/3$.

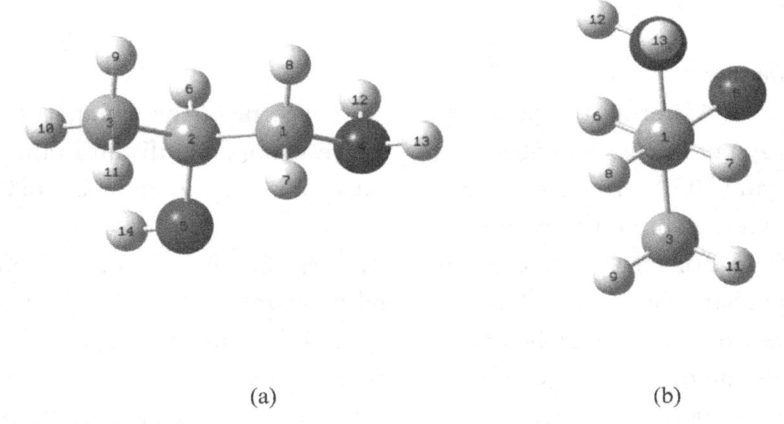

Fig. 7. 2 (a) The structure of (S)-1-amino-2-propanol. 1, 2, 3 are C, 4 is N, 5 is O and the rest are H atoms. (b) The molecular orientation of (S)-1-amino-2-propanol with the O—H bond pointing into the paper which is defined as the *Z*-axis. H14, N4, C2 and H10 are respectively behind O5, H13, C1 and C3.

Table 7. 4 The symmetry coordinates in terms of the internal coordinates. *r* denotes the bond stretching coordinate.

Symmetry coordinate	Expression in internal coordinates
S_1	r_{C1C2}
S_2	r_{C2C3}
S_3	r_{C1N4}
S_4	r_{C2O5}

S_5	r_{C2H6}
S_6	$(r_{C1H7}+r_{C1H8})/\sqrt{2}$
S_7	$(r_{C1H7}-r_{C1H8})/\sqrt{2}$
S_8	$(r_{C3H9}+r_{C3H10}+r_{C3H11})/\sqrt{3}$
S_9	$(2r_{C3H9}-r_{C3H10}-r_{C3H11})/\sqrt{6}$
S_{10}	$(r_{C3H10}-r_{C3H11})/\sqrt{2}$
S_{11}	$(r_{N4H12}+r_{N4H13})/\sqrt{2}$
S_{12}	$(r_{N4H12}-r_{N4H13})/\sqrt{2}$
S_{13}	r_{O5H14}
S_{16}	$C1H_2$ rocking
S_{27}	$C3H_3$ rocking
S_{32}	C1 out of NH_2 planewagging

Table 7.5 The Raman wavenumbers, Raman and ROA intensities (under 532 nm excitation, ICPu/SCPu (180°) configuration), PED (only those larger than 10 are shown) and ROA parameters, $\beta(G')^2$, $\beta(A)^2/3$ (10^4 Å^5/amu) for (S)-1-amino-2-propanol.

Wavenumber	Raman intensity	ROA intensity	PED	$\beta(G')^2$	$\beta(A)^2/3$
3822.40	474.3	228.9	S13(100)	2.345	0.0398
3577.47	341.2	8.6	S12(98)	0.6018	−0.5116
3492.14	674.2	−352.7	S11(120) S32(10)	−4.0488	0.3743
3104.42	237.9	427.5	S9(45) S10(53)	3.086	1.3679
3096.20	317.9	404.2	S9(53) S10(46)	3.5532	0.6581
3075.48	240.9	−1292.6	S6(30) S7(65)	−10.1911	−3.2739
3025.91	857.5	47.9	S8(99)	0.4116	0.0880

2974.66	664.3	−167.2	S5(92)	0.5615	−2.3040
2966.55	366.5	72.3	S6(63) S7(33)	−0.4749	1.2289
1136.95	13.8	18.6	S1(14) S4(37) S27(18)	0.1685	0.0254
1096.75	37.3	229.7	S2(27) S3(27)	2.2368	0.1569
1080.15	5.9	−137.2	S2(18) S3(60) S27(10)	−1.3826	−0.0466
932.87	10.2	−69.6	S1(15) S2(12) S16(18) S27(34) S32(11)	−0.5299	−0.1952
917.37	9.5	−138.7	S4(40) S16(27) S32(11)	−1.2647	−0.1804
844.02	23.3	35.9	S1(15) S2(11) S4(11) S28(22) S32(17)	0.3943	−0.02
780.7	25.8	−29.5	S1(11) S16(14) S32(92)	−0.3178	0.0101

From Table 7.5, we see that bond stretching motions are mostly coupled with the bending motions such as S16 (CH_2 rocking), S27 (CH_3 rocking) and S32 (C1—NH_2 out-of-plane wagging). Also noted is that the C—H stretching modes are very localized. Hence, for the elucidation of the bond polarizabilities, we may have two separated domains: the C—H moiety, (S5, S6, S7, S8, S9, S10, S11, S12, S13) and the molecular backbone moiety, (S1, S2, S3, S4, S16, S27, S32). For the former case, the criteria for the phase determination are that (1) the bond polarizabilities of S5, S6, S8, S11, S13 (of single bond coordinate or symmetric coordinate) are positive and (2) the bond polarizability of the symmetric coordinate is larger than that of its associate anti-symmetric coordinate in magnitude, i. e., S6 > S7; S8 > S9, S10; S11 > S12 (we mean their bond polarizabilities). With these criteria, we are left with 8 phase sets. For the latter case, the criteria are that (1) the bond polarizabilities of C1—C2, C2—C3, C—N and C—O are positive, (2) those of C1—C2 and C2—C3 are close to each other and (3) those of the bending coordinates are not too larger than those of the stretching coordinates. (In fact, these conditions are rather loose.) Luckily, we are then left with one phase set. The elucidated bond polarizabilities are shown in Fig. 7. 3(a). We note that, for the possible 8 phase sets, the indeterminate is only reflected on the two N—H bonds. In Fig. 7. 3(a), the bond electronic densities of the ground state are also shown which are calculated from the output data of Gaussian 09W relating to the *condensed to atom* charge densities.

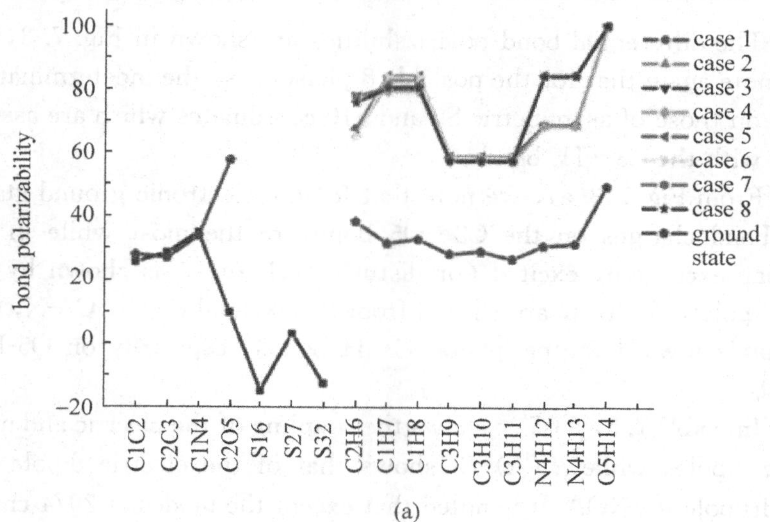

(a)

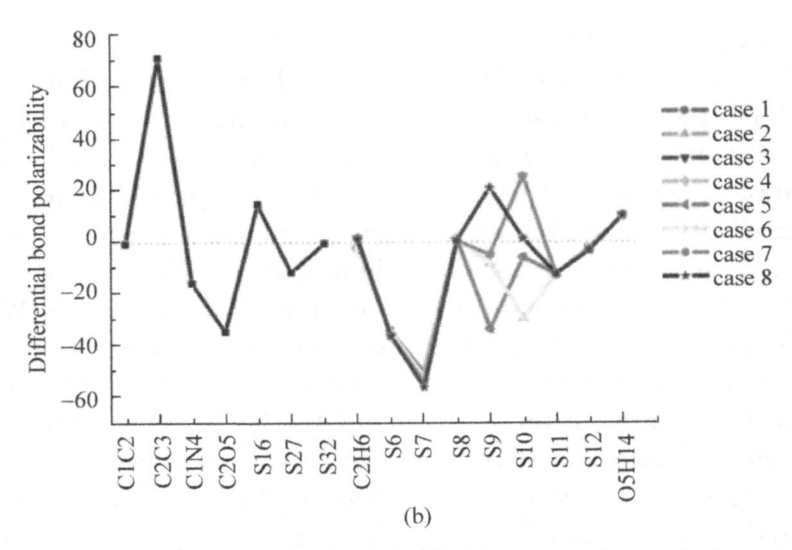

(b)

Fig. 7. 3 (a) The relative bond polarizabilities corresponding to the 8 possible phase sets and the bond electronic densities of the ground state (denoted by 'ground state' with a pentagon symbol). The bond polarizability of O5H14 is set to 100. The bond electronic densities are augmented by a factor of 10 for convenience. (b) The relative differential bond polarizabilities corresponding to the 8 possible phase sets. The value of O5H14 is set to 10.

The differential bond polarizabilities are shown in Fig. 7. 3(b). We note again that for the possible 8 phase sets, the indeterminate is only on those of asymmetric S9 and S10 coordinates which are associated with the C3—H_3 bonds.

From Fig. 7. 3(a), we note that for the electronic ground state, the bond charges on the C2—O5 bond are the most while in the Raman excitation, excited (or disturbed) charges, as shown by the bond polarizabilities, are shifted from the skeletal C—C, C—N, C—O bonds toward the peripheral C—H bonds, especially on O5-H14 bond.

In Table 7. 5, $\beta(G')^2$ shows the coupling of the electric and magnetic dipoles while $\beta(A)^2/3$ shows that of the electric dipole and quadrupole for ROA. It is noted that except the modes at 2974 cm^{-1},

2966 cm^{-1} (for them, these two ROA parameters are of opposite signs and their sums are much less than those of the other modes), all the modes are primarily operated by the coupling mechanism of the electric and magnetic dipoles. Therefore, we can try to interpret the physical significance of the differential bond polarizabilities shown in Fig. 7. 3(b) by the classical $\boldsymbol{m} \cdot \boldsymbol{\mu}$ mechanism/physical picture as demonstrated below. (Note that these quantities are due to the disturbed charges in the Raman process, *not* the permanent ones or those like in the infrared process.)

(1) The modes at 3075 cm^{-1} and 2966 cm^{-1} are primarily composed of S6 and S7 coordinates. Their ROA intensities are of opposite signs with that of 3075 cm^{-1} significantly larger and negative. Therefore, the ROA signatures of S6 and S7 are determined by the ROA intensity at 3075 cm^{-1} and both are negative. This is consistent with their differential bond polarizabilities. Hence, we are confident in the phase sets determined. (The inconsistency for 2966 cm^{-1} is due to that if a mode is composed of several bond coordinates with comparable components, then the differential bond polarizability signs of these bond coordinates are not related to the mode ROA signature in a simple and straight way. This can be significant for small ROA peak.)

(2) The 2974cm^{-1} mode is mainly due to the C2—H6 stretching motion. Its two ROA parameters, $\beta(G')^2$, $\beta(A)^2/3$ are of opposite signs and their sum is small. This is consistent to its small differential bond polarizability.

(3) In Raman excitation, charges are more shifted on the O—H bond as pointed out previously based on the bond polarizabilities (see Fig. 7. 3(a)). We suppose that the magnetic field operating in ROA for this molecule is dominantly induced by the vibrationally induced charge motion on the O—H bond. If we take the orientation of the O—H bond as the *Z*-axis, then the induced magnetic field is more or less lying in the *X*-*Y* plane (we mean, of course, all the planes parallel to it). Hence, for those bonds that are parallel to the *Z*-axis, their $\boldsymbol{m} \cdot \boldsymbol{\mu}$ couplings between their vibrating electric dipoles and the magnetic field originating from the O—H bond will be vanishing. Instead, those bonds that are more vertical to the *Z*-axis or lie in the *X*-*Y* plane

will have larger $\boldsymbol{m} \cdot \boldsymbol{\mu}$ coupling strength and possess larger differential bond polarizabilities. The molecular orientation with the O—H bond pointing into the paper, defined as the *Z*-axis, is shown in Fig. 7. 2 (b).

With the above assessment, we note indeed that the differential bond polarizability of C1—C2, whose orientation is almost parallel to the Z-axis, is vanishing while those of C2—C3, C1—N4 and C2—O5, which possess larger projections in the *X*-*Y* plane, are eminent. The differential bond polarizability of C2—H6 is also vanishing. The projection of this bond in the *X*-*Y* plane is less than the above bonds, hence its $\boldsymbol{m} \cdot \boldsymbol{\mu}$ will not be significant. (Note this bond is mainly associated with the mode at 2974 cm^{-1} which possesses very small sum of $\beta(G')^2$ and $\beta(A)^2/3$ as was noted in paragraph(2). This leads to its small ROA activity.) The dipole direction of antisymmetric S7 is in the *X*-*Y* plane, while that of symmetric S6 is with an angle of 40°to the *X*-*Y* plane. Hence the former possesses larger differential bond polarizability.

The angles of the dipole directions to the *X*-*Y* plane of S8, S9 and S10 are respectively 15°, 23°, 62°. Hence, it is expected that their differential bond polarizabilities are in the decreasing order. This apparently contradicts with the vanishing differential bond polarizability of S8. The vanishing differential bond polarizability of S8 may be caused by other factors. One factor is that in ROA, the chirality will be more eminent for the off-diagonal polarizability tensor elements. S8 is symmetric with more diagonal polarizability elements and hence, its differential bond polarizability can be less than those of asymmetric S9 and S10 which are with more off-diagonal polarizability elements. For this, we note that the previous observation that the differential bond polarizability of anti-symmetric S7 is larger than that of its counterpart, the symmetric S6, is in consistence with this assertion.

The ROA signatures of the modes around 3104 cm^{-1} and 3096 cm^{-1} are positive. These two modes are mainly due to S9 and S10. Their differential bond polarizabilities could also be positive. This will help in the further nail-down of the indeterminate phase sets. This leads to the solutions of case 4 and case 8(in fact, they are almost identical). In this way, we have that the differential bond polarizability of S9 is larger than that of S10 as anticipated from our assertion (see the previous paragraph) based on the bond (or its dipole) orientation. The differential bond polarizability of S10 is quite small since the angle of its dipole orientation to the *X-Y* plane is quite large (62°). That the differential bond polarizabilities of S9, S10 are less than those of S6 and S7 could be associated with the fact that in the Raman excitation, the charges (evidenced by bond polarizability and are associated with $\boldsymbol{\mu}$) on C3—H are less than those on C1—H as shown in Fig. 7. 3(a).

The angles of the dipole orientation to the *X-Y* plane for S11 and S12 are quite large (the former is 43. 5° and the latter is 43. 6°). This leads to their small differential bond polarizabilities. If this is also associated with their less charges (on the N—H bond) during Raman excitation, then we prefer the solutions corresponding to case 1 to case 4 (for which the bond polarizabilities are quite identical) as shown in Fig. 7. 3(a). Together with the previous assertion of the choice of case 4 and case 8, the true phase set then can be case 4.

In summary, for this (S)-1-amino-2-propanol case, we found that during Raman excitation, charges are more aggregated on its O—H bond. The ROA mechanism for this molecule is mainly operated by the coupling between the vibrationally induced electric and magnetic dipoles. Hence, classically, this ROA mechanism predicts that those bonds less parallel to the O—H bond will acquire more chiral activity. This is confirmed by our differential bond polarizability analysis on the ROA intensities that C2—C3, C1—N4, C2—O5 bonds and antisymmetric mode S7, of which electric dipoles are less parallel to the O—H bond, acquire more chiral activity while C1—C2, C2—H6 bonds and symmetric mode S6, of which electric dipoles are more parallel to the O—H bond, show the opposite character.

Comments

It is demonstrated that ROA spectrum does contain much stereostructural information. The issue is how to retrieve it. If one sticks only to quantal algorithm, he or she may be able to have accurate ROA data. However, the physics or the physical picture to him or her may be just a maze. In this sense, the classical or semiclassical way to comprehend ROA intensities is indispensable. In this aspect, we have demonstrated that bond polarizability algorithm is an adequate means to serve this purpose.

7.3 The case of (2R, 3R)-(−)-2, 3-Butanediol

The structure of(2R, 3R)-(−)-2, 3-butanediol(2, 3-BDO)can be optimized by Gaussian09W software(B3LYP, 6-31G**)with either C_1 or C_2 symmetry. We will treat them separately. Shown in Fig. 7. 4 are their structures. Shown in Fig. 7. 5 are its Raman and ROA spectra by 532 nm excitation. In Fig. 7. 5, those marked by * are employed for retrieving bond polarizabilities and differential bond polarizabilities under C_1 symmetry. In(a), there are 11 *'s and in(b), there are 4 *'s. + and ^ are those for retrieving bond polarizabilities and differential bond polarizabilities of the symmetric and anti-symmetric coordinates under C_2 symmetry. For the symmetric coordinates, there are 15 +'s in(a)and 10 +'s in(b)(other 5 +'s are in the high wavenumber region, not shown.). For the anti-symmetric coordinates, there are 10 ^'s in(a)(other 5 ^'s are too weak to be labeled)and 8 ^'s in(b)(other 2 ^'s are missing)

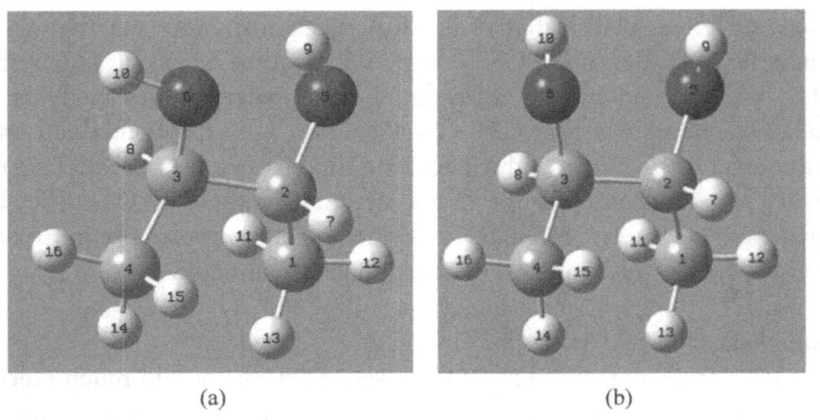

Fig. 7. 4 The structures of 2, 3-BDO with(a)C_1, (b)C_2 symmetry. 1-4 are C atom, 5-6 are O atom and the rest are H atom.

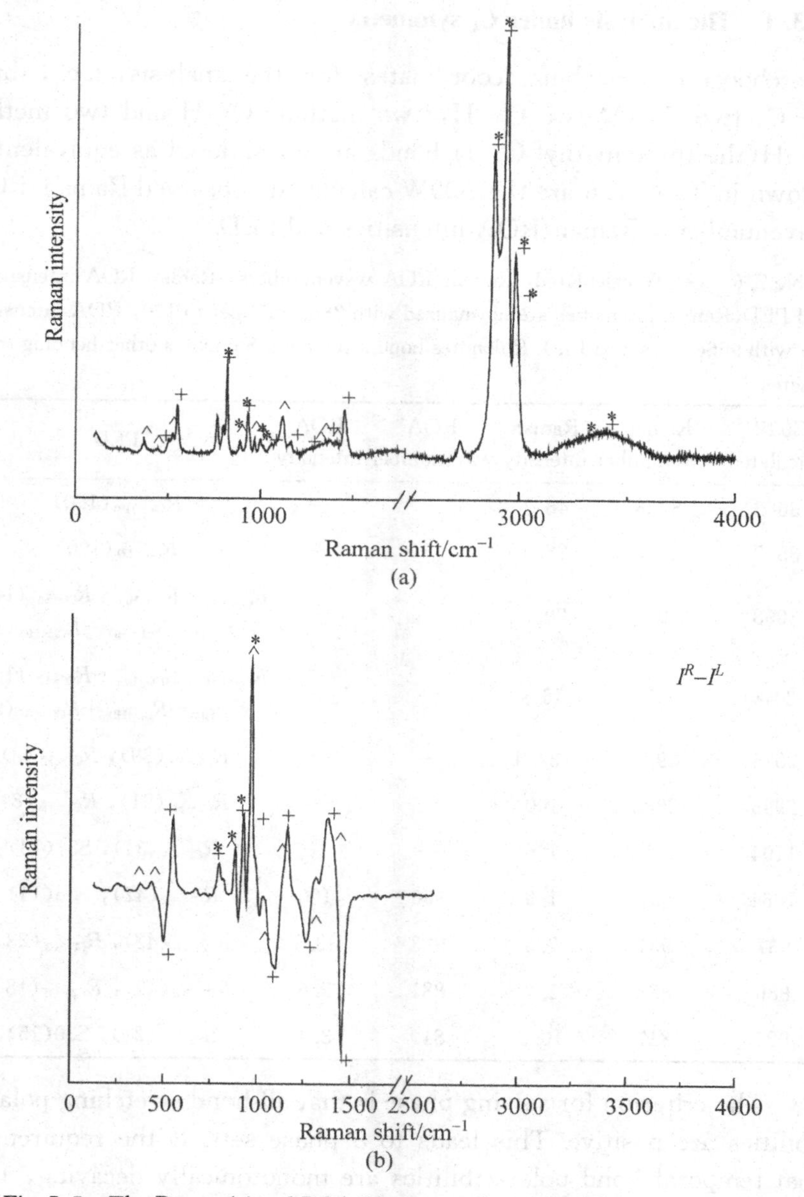

Fig. 7.5 The Raman(a) and ROA(b) spectra of 2, 3-BDO by 532 nm excitation. See text for the meanings of ∗, + and ^.

7.3.1 The analysis under C_1 symmetry

We choose 11 stretching coordinates for the analysis, i.e., three C—C, two C—O, two O—H, two methine C—H and two methyl C—H(the three methyl C—H bonds are consi-dered as equivalent). Shown in Table 7.6 are the G09W calculated, observed Raman/ROA wavenumbers, Raman/ROA intensities and PED.

Table 7.6 G09W calculated, Raman/ROA wavenumbers, Raman/ROA intensities, and PED(Raman intensities are normalized with 2882cm^{-1} set to 100, ROA intensities are with 966cm^{-1} set to 100). *R* denotes bond stretching. *S* denotes other bending coordinates

G09W calculated	Raman wavenumber	Raman intensity	ROA wavenumber	ROA intensity	PED
3663	3426	48.7	—	—	R_{O6-H10}(100)
3597	3289	25.4	—	—	R_{O5-H9}(100)
3058	2989	20.5	—	—	$R_{C1-H13}+R_{C1-H11}+R_{C1-H12}$(144), $R_{C1-H13}+R_{C1-H11}-R_{C1-H12}$(29)
3048	2977	16.8	—	—	$R_{C4-H14}+R_{C4-H15}+R_{C4-H16}$(142), $R_{C4-H14}-R_{C4-H15}+R_{C4-H16}$(20)
3025	2934	130.4	—	—	R_{C3-H8}(89), R_{C2-H7}(8)
2995	2882	100	—	—	R_{C2-H7}(91), R_{C3-H8}(8)
1104	1013	1.6	—	—	R_{C3-C4}(31), S37(13)
1064	997	1.9	966	100	R_{C1-C2}(42), S36(14)
937	931	7.5	923	33.1	R_{C2-O5}(42), R_{C2-C3}(23)
886	888	1.7	881	5.6	R_{C3-O6}(49), R_{C2-O5}(18)
825	817	10.2	819	3.4	R_{C2-C3}(21), S30(15)

The criterion for solving phase is that all bond stretching polarizabilities are positive. This leads to 8 phase sets. If the requirement that temporal bond polarizabilities are monotonically decaying, then only one phase set is left. Shown in Fig. 7. 6 are the bond polarizabilities at the initial Raman excitation and after relaxation (1 ps). The bond densities by EHMO are also shown therein. From Fig. 7. 6, it is noted that the bond polarizabilities of C2—H7, C3—H8; C1—H13, C4—H14; O5—H9, O6—H10 are more eminent while those of the skeletal C1—C2 and C3—C4 bonds are small. This shows that during Raman excitation, charges are flown to the periphery. After relaxation, the bond polarizabilities and bond densities are consistent.

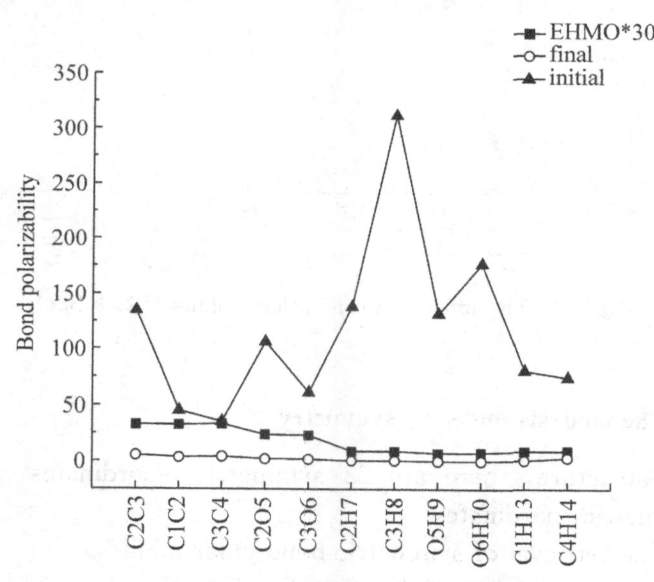

Fig. 7. 6 The bond polarizabilities at the initial Raman excitation and after relaxation(1 ps). The bond densities by EHMO are also shown therein.

Shown in Fig. 7. 7 are the differential bond polarizabilities. From the figure, it is noted that those of C1—C2 and C3—C4 are negative while those of C2—O5 and C3—O6 which are on the other side of the plane formed by H7—C2—C1—H8 are positive. The sign of that of C2—C3 is also positive. This shows that ROA asymmetry around chiral C2 and C3 atoms is very definite.

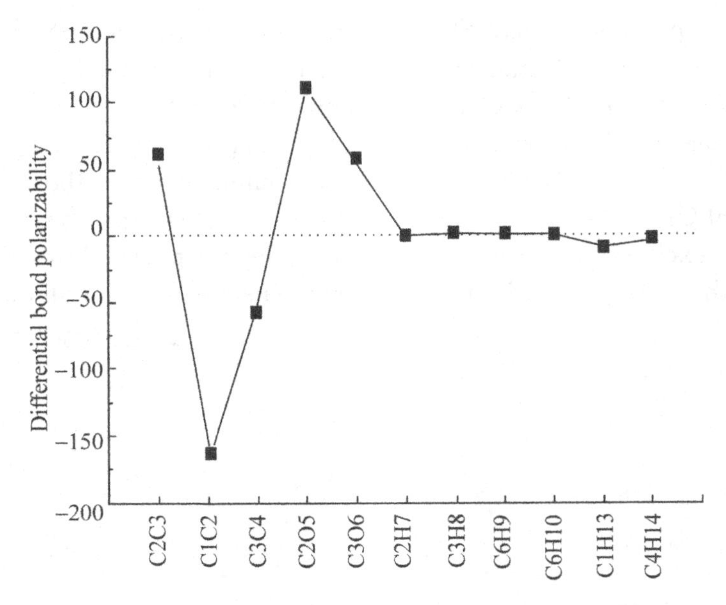

Fig. 7. 7 The differential bond polarizabilities of 2, 3-BDO.

7. 3. 2 The analysis under C_2 symmetry

For this structure, there are 22 symmetric coordinates and 20 anti-symmetric coordinates.

1. The retrieval of symmetric bond polarizabilities

Symmetric modes are labeled in Fig. 7. 5. We chose 15 modes (8 for stretchings, 7 for bendings with significant intensities). Table 7. 7 shows their Raman wavenumbers, intensities, ROA intensities and PED. We treat the 6 methyl C—H bonds as equivalent, denoted by C1—H11. Hence, only the mode at 2934cm^{-1} is needed for the methyl C—H bonds. Then, 13 modes are required corresponding to C1—C2, C2—C3, C2—O5, C2—H7, O5—H9, C1—H11 and 7 bending coordinates with larger ROA intensities. The criteria for phase solution are that all 6 bond stretching polarizabilities are positive and decay in a monotonic way during relaxation. This leads to 2 phase sets. Fig. 7. 8

shows the symmetric bond polarizabilities. It is noted that those of C2—H7, O5—H9 , C1—H11 are larger. This is consistent with that under C_1 structural analysis.

Table 7. 7 G09W calculated, Raman wavenumbers, Raman/ROA intensities, and PED for the symmetric modes (Raman intensities are normalized with 2882cm^{-1} set to 100, ROA intensities are with 966cm^{-1} set to 100). *R* denotes bond stretching. *S* denotes other bending coordinates.

G09W calculated	Raman wavenumber	Raman intensity	ROA intensity	PED
3819	3426	48.7	—	$R_{O5-H9}+R_{O6-H10}$ (100)
3131	2989	20.5	—	$2R_{C1-H11}-R_{C1-H12}-R_{C1-H13}+2R_{C4-H14}-R_{C4-H15}-R_{C4-H16}$ (97)
3115	2976	16.8	—	$R_{C1-H11}-R_{C1-H12}+R_{C4-H14}-R_{C4-H15}$ (93)
3043	2934	130.4	—	$R_{C1-H11}+R_{C1-H12}+R_{C1-H13}+R_{C4-H14}+R_{C4-H15}+R_{C4-H16}$ (96)
2991	2882	100.0	—	$R_{C2-H7}+R_{C3-H8}$ (100)
1482	1454	11.7	−136.5	S28(48), S30(48)
1385	1375	4.0	37.9	S16(69), S26(13)
1263	1260	2.2	−26.4	S36(41), S18(27)
1176	1163	1.7	55.8	R_{C2-C3} (32), S34(16)
1105	1077	7.7	−77.0	$R_{C2-O5}+R_{C3-O6}$ (38), S32(18)
1049	997	1.9	9.6	$R_{C1-C2}+R_{C3-C4}$ (55), S36(21)
926	931	7.5	33.1	S32(45), $R_{C2-O5}+R_{C3-O6}$ (31)
814	817	10.2	19.6	S34(38), R_{C2-C3} (22)
542	550	8.7	47.1	S24(44), S22(21)
493	504	2.3	−42.8	S41(90), S24(7)

Shown in Fig. 7. 9 are the symmetric differential bond polarizabilities. In the figure, we note that the signs of those of C2—C3 and C2—O5 are the same while that of C1—C2 is opposite. This is just consistent with C_1 structural analysis.

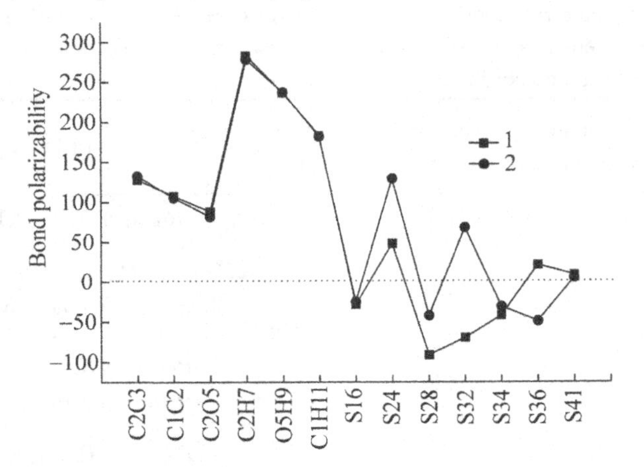

Fig. 7. 8 The symmetric bond polarizabilities of 2, 3-BDO. S refers to the bending coordinates.

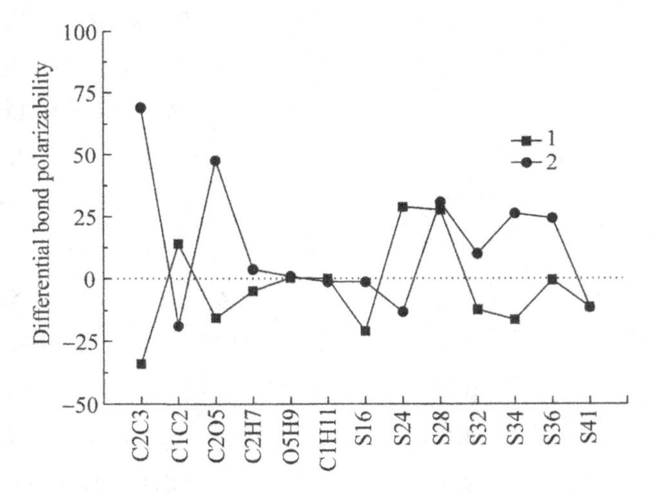

Fig. 7. 9 The symmetric differential bond polarizabilities of 2, 3-BDO. S refers to the bending coordinates.

2. The retrieval of anti-symmetric bond polarizabilities

Anti-symmetric modes are labeled in Fig. 7. 5. We chose 15 modes (7 for stretchings and 8 for bendings with significant Raman/ROA intensities). Table 7. 8 shows their Raman wavenumbers, intensities, ROA intensities and PED. Since both the symmetric and anti-symmetric mode intensities are based on the mode intensity at 2882cm^{-1} for Raman and that at $966\ \text{cm}^{-1}$ for ROA, their bond polarizabilities and differential bond polarizabilities are comparable. The criteria for solving phases are that all those for C—C, C—O, O—H and C—H are positive and decay in a monotonic way during relaxation. This leads to 6 phase sets. Shown in Fig. 7. 10 are the anti-symmetric bond polarizabilities. These six sets of bond polarizabilities are quite consistent. Noticeable is that of C1—C2 is smaller than that of C2—O5. Those of S21(C3C2O5, C3C2O6 deformation) and S23(C1C2C3, C4C3C2 deformation) are large and of opposite signs.

Table 7. 8 G09W calculated, Raman wavenumbers, Raman/ROA intensities, and PED for the anti-symmetric modes (Raman intensities are normalized with 2882 cm^{-1} set to 100, ROA intensities are with 966 cm^{-1} set to 100). *R* denotes bond stretching. *S* denotes other bending coordinates.

G09W calculated	Raman wavenumber	Raman intensity	ROA intensity	PED
3817	—	0	—	$R_{O5-H9}-R_{O6-H10}$ (100)
3125	—	0	—	$R_{C1-H11}-R_{C1-H12}-R_{C4-H14}+R_{C4-H15}$ (99)
3115	—	0	—	$2R_{C1-H11}-R_{C1-H12}-R_{C1-H13}-2R_{C4-H14}+R_{C4-H15}+R_{C4-H16}$ (95)
3041	—	0	—	$R_{C1-H11}+R_{C1-H12}+R_{C1-H13}-R_{C4-H14}-R_{C4-H15}-R_{C4-H16}$ (97)
3006	—	0	—	$R_{C2-H7}-R_{C3-H8}$ (98)
1414	1405	2. 1	114. 5	S19(39), S27(29)
1375	1348	3. 5	—	S27(68), S19(22)

1295	1282	0. 8	—26. 2	S17(66), S35(9)
1130	1118	10. 4	23. 8	$R_{C2-O5}-R_{C3-O6}$(43), S33(21)
1044	1013	1. 6	—11. 4	$R_{C1-C2}-R_{C3-C4}$(40), S37(25)
972	966	1. 2	100. 0	S35(61), S17(19)
895	888	1. 7	5. 6	$R_{C2-O5}-R_{C3-O6}$(46), S33(28)
502	525	2. 5	—	S21(70), $R_{C1-C2}-R_{C3-C4}$(10)
421	441	0. 9	11. 6	S25(89), S33(20)
357	372	3. 3	6. 1	S23(101), S21(27)

Shown in Fig. 7. 11 are the anti-symmetric differential bond polarizabilities. It is evident that the sign for C1—C2 is negative and that for C2—O5 is positive. This is consistent with the C_1 structural analysis.

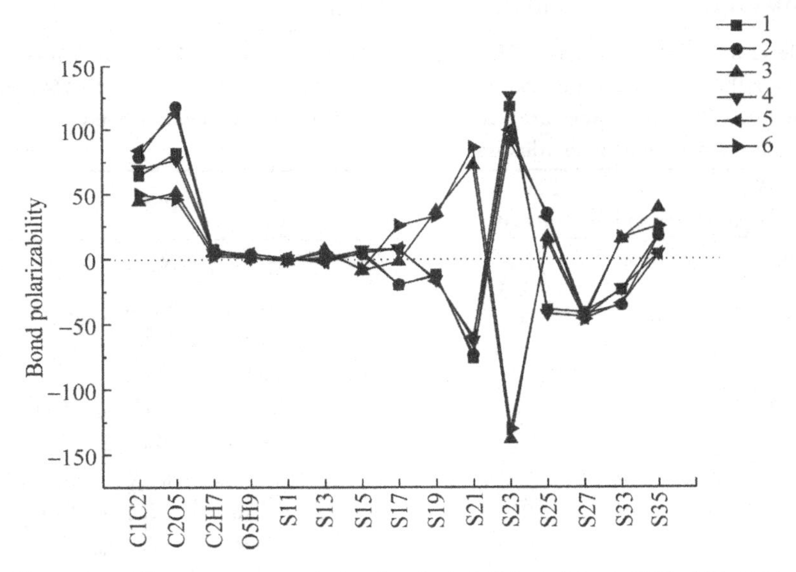

Fig. 7. 10 The anti-symmetric bond polarizabilities of 2, 3-BDO. S11, S13, S15 are related to the stretchings of the methyl group and the rest are the bending coordinates.

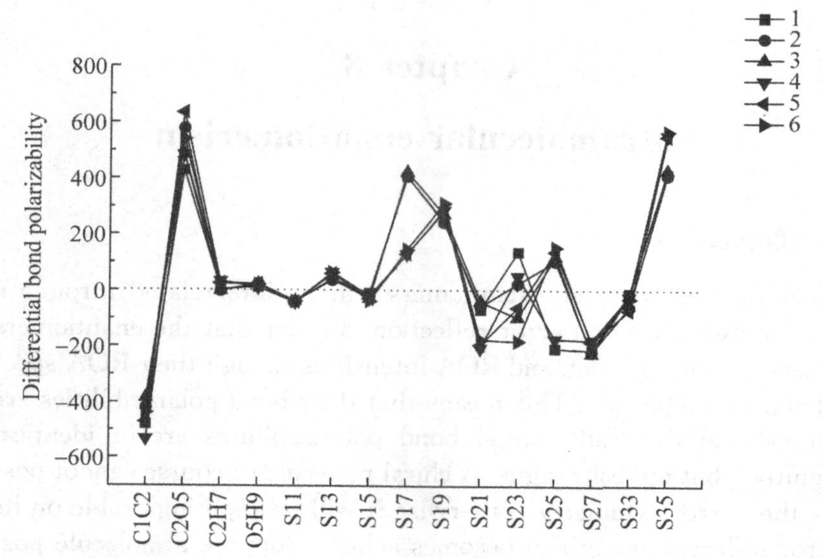

Fig. 7. 11 The anti-symmetric differential bond polarizabilities of 2, 3-BDO. S11, S13 and S15 are related to the stretchings of the methyl group and the rest are the bending coordinates.

Comments

(1) The consistency of the results by C_1 and C_2 structural analysis demonstrates that the bond polarizability algorithm is quite independent of the arbitration in the structural supposition.

(2) It is noticed that for Raman effect, it is the symmetric coordinates that are more enhanced while for ROA effect, it is the anti-symmetric coordinates that are more enhanced. For this, readers are referred to Section 6. 5 and 7. 1.

References

[1] Barron L D. Molecular Light Scattering and Optical Activity, 2nd edition University Press, Cambridge, 2004.

[2] Nafie L A. Vibrational Optical Activity: Principles and Applications, John Wiley & Sons Ltd. , West Sussex, 2011.

Chapter 8
Intramolecular enantiomerism

8.1 Background

The chiral right and left enantiomers can be interrelated through a mirror reflection. The mirror reflection is strict that the enantiomers possess identical Raman and ROA intensities though their ROA spectral signs are opposite. This means that their bond polarizabilities are identical and their differential bond polarizabilities are of identical magnitude but opposite signs. A chiral molecule of course cannot possess the mirror symmetry, otherwise it will be superimposable on its mirror reflected entity and becomes achiral. Suppose a molecule possesses a mirror symmetry and we break the mirror symmetry by the addition/replacement of a functional group. Then the mirror symmetry is lost and the two parts in the molecule which were originally mirror images to each other are no longer superimposable. Thus, their bond polarizabilities and differential bond polarizabilities are irrelevant to each other, including the signs of the differential bond polarizabilities. This is justifiable since the mirror symmetry breaking by the addition/replacement of a functional group is not a small perturbation. Rather unexpected, we will demonstrate in this chapter that the situation is not so simple. For some chiral molecules that will be shown in this chapter, the two parts which were otherwise related by the mirror reflection, do possess counter signs for their differential bond polarizabilities. This means that due to the breaking of mirror symmetry, the molecule now acquires chirality, however, the symmetry breaking is not strong enough that it still possesses the property very similar to that between the right and left enantiomers as far as the signs of the differential bond polarizabilities are concerned. This is somewhat out of expectation. We call it the intra-molecular enantiomerism.

8.2 The case of (R)-(+)-Limonene

The structure of (R)-Limonene((R)-1-Methyl-4-isopropenyl-1-cyclohexene)is optimized by DFT B3LYP at the level of 6-311g by Gaussian09W. Its structure and atomic numberings are shown in Fig. 8.1(a). Its Raman and ROA spectra are shown in Fig. 8.2. We will pay our most attention to its skeletal C—C stretching and bending motions which are below 2000 cm^{-1}. Those marked by * in Fig. 8.2 are the peaks employed for the elucidation of the bond polarizabilities and differential bond polarizabilities(their choices are explained in the next section). Their Raman shifts, relative Raman and ROA intensities and PED are shown in Table 8.1. The related symmetry coordinates are shown in Table 8.2.

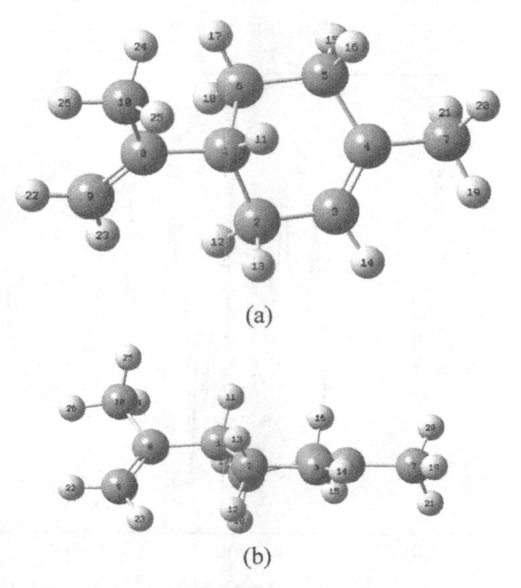

Fig. 8.1 (a)The structure of (R)-limonene and its atomic numberings. 1~10 are the C atom and the rest are the H atom. (b)The side view showing that the six membered ring is roughly planar except C1 atom. Note that the vertical mirror along C1 and C4 atoms and the mirror along the mid-points of C5—C6 and C2—C3 bonds play the role of enantiomerism.

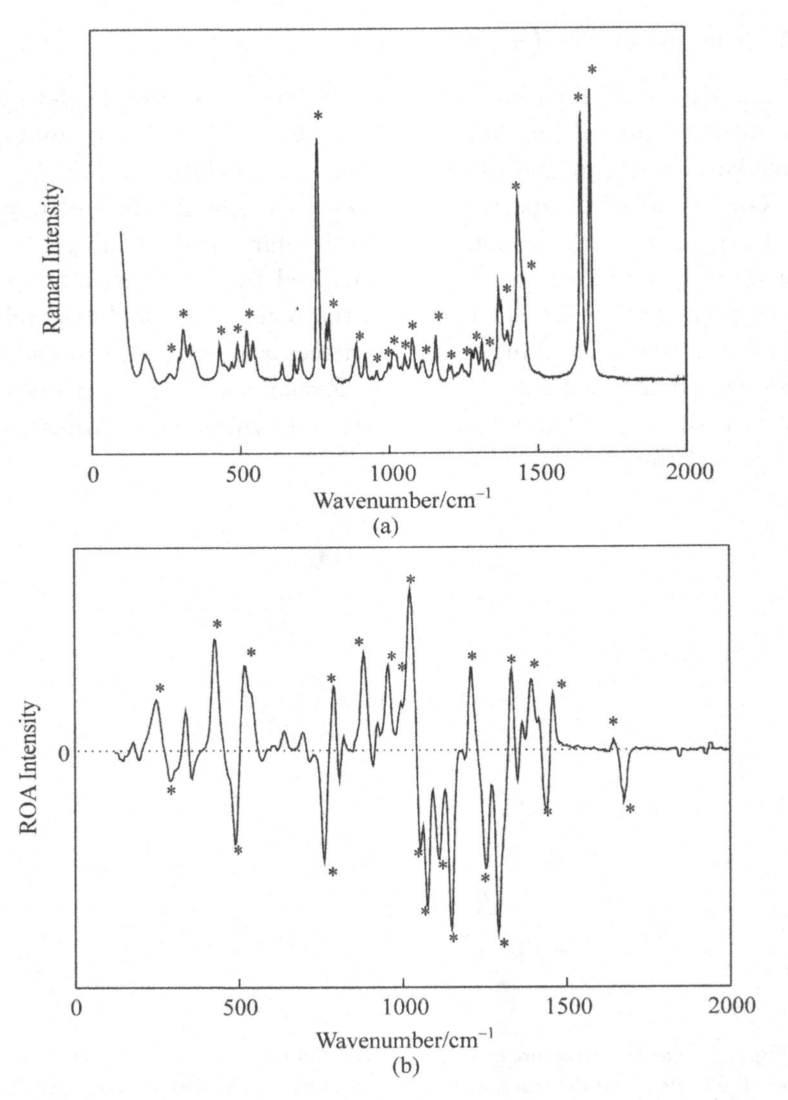

Fig. 8. 2 The Raman (a) and ROA (b) spectra of (R)-limonene. Those marked by * are employed for the retrieval of bond polarizabilities and differential bond polarizabilities.

Table 8.1 The calculated and observed Raman wavenumbers, observed relative Raman and ROA intensities and PED (only those larger than 10 are listed).

Calculated wavenumber	Raman wavenumber	Raman intensity	ROA intensity	PED
1677.6	1678.1	84.0	−21.6	S_5(77)
1650.3	1645.6	96.3	4.1	S_9(74), S_{55}(18)
1470.8	1452.8	46.5	26.3	S_{36}(28), S_{58}(27)
1435.9	1435.1	101.4	−35.6	S_{55}(65), S_{59}(11)
1408.7	1394.6	45.1	41.9	S_{31}(103)
1339.7	1331.5	3.0	26.3	S_{52}(37), S_{42}(23)
1312.3	1290.1	12.1	−80.7	S_{42}(19), S_{37}(15)
1265.1	1259.9	1.5	−57.1	S_{49}(31), S_{41}(19)
1222.5	1208.0	5.2	44.5	S_{49}(17), S_{37}(15)
1153.0	1147.8	3.4	−67.2	S_7(32), S_6(11)
1106.0	1112.5	12.9	−60.1	S_2(34), S_3(16)
1065.1	1078.8	19.9	−100.0	S_{60}(47), S_{61}(15)
1048.3	1053.2	8.3	−31.2	S_2(23), S_{34}(13)
1021.6	1021.0	19.3	81.1	S_4(32), S_{34}(14)
983.8	996.3	6.9	18.4	S_8(19), S_{56}(16)
949.8	957.0	3.9	44.5	S_{51}(14), S_{39}(11)
875.2	888.6	19.1	51.5	S_{10}(36), S_1(28)
800.4	789.0	6.0	37.8	S_{39}(33), S_{43}(20)
748.0	760.4	85.5	−44.6	S_6(30), S_{28}(10)
505.9	522.3	17.3	81.6	S_{29}(36), S_{64}(29)
494.0	491.0	18.6	−60.4	S_{54}(28), S_{47}(11)
434.4	429.3	11.7	69.5	S_{62}(59), S_{70}(28)
299.8	293.1	3.3	−47.4	S_{28}(24), S_{47}(15)
260.4	264.2	5.2	70.5	S_{72}(23), S_{70}(16)

Table 8. 2 The local symmetry coordinates of (R)-limonene. r and α denote the bond stretching and bending(angle) coordinates. β and τ denote out-of-plane and twist coordinates.

Definition	Description
$S_1 = r(C1-C2)$	C1—C2 stretching
$S_2 = r(C1-C6)$	C1—C6 stretching
$S_3 = r(C2-C3)$	C2—C3 stretching
$S_4 = r(C5-C6)$	C5—C6 stretching
$S_5 = r(C3-C4)$	C3—C4 stretching
$S_6 = r(C4-C5)$	C4—C5 stretching
$S_7 = r(C4-C7)$	C4—C7 stretching
$S_8 = r(C1-C8)$	C1—C8 stretching
$S_9 = r(C8-C9)$	C8—C9 stretching
$S_{10} = r(C8-C10)$	C8—C10 stretching
$S_{28} = 12^{-1/2}[2\alpha(C6-C1-C2) - \alpha(C1-C2-C3) - \alpha(C2-C3-C4) + 2\alpha(C3-C4-C5) - \alpha(C4-C5-C6) - \alpha(C5-C6-C1)]$	ring asym def
$S_{29} = 2^{-1}[\alpha(C1-C2-C3) - \alpha(C2-C3-C4) + \alpha(C4-C5-C6) - \alpha(C5-C6-C1)]$	ring asym def'
$S_{31} = 6^{-1/2}[\alpha(H20-C7-H21) + \alpha(H19-C7-H21) + \alpha(H19-C7-H20) - \alpha(C4-C7-H19) - \alpha(C4-C7-H20) - \alpha(C4-C7-H21)]$	C7—H_3 sym def
$S_{34} = 6^{-1/2}[2\alpha(C4-C7-H19) - \alpha(C4-C7-H20) - \alpha(C4-C7-H21)]$	C7—H_3 rocking
$S_{36} = \alpha(H15-C5-H16)$	C5—H_2 scissoring
$S_{37} = 2^{-1}[\alpha(C4-C5-H16) - \alpha(C4-C5-H15) + \alpha(C6-C5-H15) - \alpha(C6-C5-H16)]$	C5—H_2 rocking
$S_{39} = 2^{-1}[\alpha(C4-C5-H16) - \alpha(C4-C5-H15) - \alpha(C6-C5-H15) + \alpha(C6-C5-H16)]$	C5—H_2 twisting
$S_{41} = 2^{-1}[\alpha(C5-C6-H17) - \alpha(C5-C6-H18) + \alpha(C1-C6-H18) - \alpha(C1-C6-H17)]$	C6—H_2 rocking

$S_{42}=2^{-1}[\alpha(C5-C6-H17)+\alpha(C5-C6-H18)$ $-\alpha(C1-C6-H18)-\alpha(C1-C6-H17)]$	C6—H_2 wagging
$S_{43}=2^{-1}[\alpha(C5-C6-H17)-\alpha(C5-C6-H18)$ $-\alpha(C1-C6-H18)+\alpha(C1-C6-H17)]$	C6—H_2 twisting
$S_{47}=17^{-1/2}[\alpha(C6-C1-C8)+4\alpha(C2-C1-C8)]$	C2—C1—C8 def
$S_{49}=2^{-1}[\alpha(C3-C2-H12)-\alpha(C3-C2-H13)$ $+\alpha(C1-C2-H13)-\alpha(C1-C2-H12)]$	C2—H_2 rocking
$S_{51}=2^{-1}[\alpha(C3-C2-H12)-\alpha(C3-C2-H13)$ $-\alpha(C1-C2-H13)+\alpha(C1-C2-H12)]$	C2—H_2 twisting
$S_{52}=2^{-1/2}[\alpha(C4-C3-H14)-\alpha(C2-C3-H14)]$	C3—H14 rocking
$S_{54}=6^{-1/2}[-\alpha(C9-C8-C10)-\alpha(C1-C8-C10)$ $+2\alpha(C1-C8-C9)]$	C1—C8—C9 def
$S_{55}=6^{-1/2}[2\alpha(H22-C9-H23)-\alpha(C8-C9-H22)$ $-\alpha(C8-C9-H23)]$	C9—H_2 sym def
$S_{56}=2^{-1/2}[\alpha(C8-C9-H22)-\alpha(C8-C9-H23)]$	C9—H_2 rocking
$S_{58}=6^{-1/2}[2\alpha(H25-C10-H26)-\alpha(H24-C10-H26)$ $-\alpha(H24-C10-H25)]$	C10—H_3 asym def
$S_{59}=2^{-1/2}[\alpha(H24-C10-H26)-\alpha(H24-C10-H25)]$	C10—H_3 asym def'
$S_{60}=6^{-1/2}[2\alpha(C8-C10-H24)-\alpha(C8-C10-H25)$ $-\alpha(C8-C10-H26)]$	C10—H_3 rocking
$S_{61}=2^{-1/2}[\alpha(C8-C10-H25)-\alpha(C8-C10-H26)]$	C10—H_3 rocking'
$S_{62}=\beta(C7-C3C4C5)$	C7—C4 wagging
$S_{64}=\beta(C10-C8C1C9)$	C10—C8 wagging
$S_{70}=6^{-1/2}[\tau(C1-C2)-\tau(C2-C3)+\tau(C3-C4)$ $-\tau(C4-C5)+\tau(C5-C6)-\tau(C6-C1)]$	ring puckering
$S_{72}=12^{-1/2}[-\tau(C1-C2)+2\tau(C2-C3)-\tau(C3-C4)$ $-\tau(C4-C5)+2\tau(C5-C6)-\tau(C6-C1)]$	ring asym torsion

We chose 10 C—C stretching coordinates (C1—C2, C2—C3, C3—C4, C4—C5, C5—C6, C6—C1, C1—C8, C4—C7, C8—C9, C8—C10) and 14 bending coordinates (S29, S31, S39, S42, S47, S49, S52, S54, S55, S56, S58, S60, S62, S70) for the Raman and ROA intensity analyses. These coordinates are more or less coupled to one another. Their associated intensities are more intense (as marked and listed in Fig. 8. 2 and Table 8. 1 respectively.). Totally, we have 24 bond coordinates and 24 corresponding Raman mode intensities. The criterion for the phase solution is that the bond polarizabilities of the 10 C—C stretching coordinates are positive and their variations (differences) are no more than 10 fold (This number is in fact not so crucial.). Then, 8 phase sets left. The corresponding bond polarizabilities are shown in Fig. 8. 3(a). We see that except those of S31, S58, S60, the rest are rather consistent despite of the 8 phase sets. We note that the bond polarizabilities of C3—C4 and C8—C9 coordinates are significantly larger than the rest C—C coordinates. This is expected since these two coordinates are the double bonds and they are richer in charges than the rest single bonds. So are larger their bond polarizabilities. The differential bond polarizabilities are shown in Fig. 8. 3(b). It is interesting to note that the variations of the differential bond polarizabilities under the 8 phase sets are very small. Hence, we have solved this phase issue despite that we have 8 inderminate phase sets.

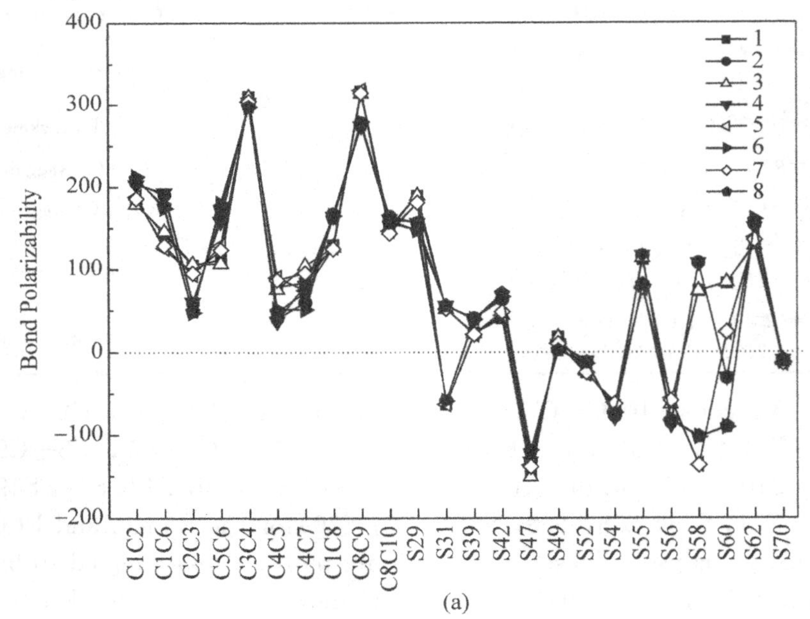

(a)

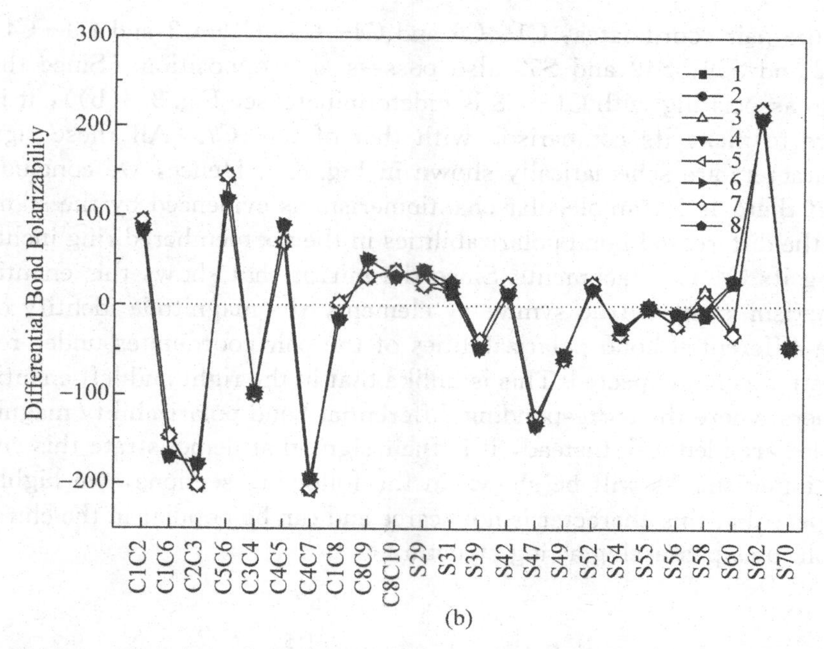

Fig. 8. 3 The relative bond polarizabilities (a) and differential bond polarizabilities (b) of (R)-limonene by the 8 indeterminate phase sets.

(R)-Limonene possesses a six membered ring structure which is roughly planar (except C1 atom). C1 is the asymmetric center. See Fig. 8. 1 (b). If we neglect the variation of the C3=C4 double bond with the rest five C—C single bonds and the rest peripheral attachments to the ring, we may consider that there is a mirror symmetry by the vertical (to the molecular plane) mirror along C1 and C4 atoms. So is a mirror symmetry by the mirror along the mid-points of C5—C6 and C2—C3 bonds (and vertical to the molecular plane).

From Fig. 8. 3 (b), it is seen that the differential bond polarizabilities of the pair coordinates: by the aforementioned mirror reflection, C1—C2 and C1—C6; C2—C3 and C5—C6; C3—C4 and C4—C5 are of opposite signs. Furthermore, those of the pair bending coordinates: S42 (C6—H_2 bending) and S49 (C2—H_2 bending); S39 (C5—H_2 bending) and S52 (C3—H bending) also show opposite signs. Along the other mirror bisecting the six membered ring, we note that those of the pair coordinates: C1—C6 and C4—C5; C1—C2 and C3—C4; S42 and S39; S49 and S52 also possess sign opposition. (Since the sign associating with C1—C8 is indeterminate(see Fig. 8. 3(b)), it is hard to make its comparison with that of C4—C7.) All these sign characters are schematically shown in Fig. 8. 4. Hence, we conclude that there is intramolecular enantiomerism as evidenced by the signs of the differential bond polarizabilities in the six membered ring including its C—H attachments. Since the mirror that shows this enantiomerism is not a true symmetry element, the magnitude identity of the differential bond polarizabilities of the pair coordinates under reflection is not expected. This is unlike that in the right and left enantiomers where the corresponding differential bond polarizability magnitudes are identical. Instead, it is their signs that demonstrate this enantiomerism. As will be shown in the following sections, we highly expect that this character is not scarce and can be popular in the chiral molecules possessing a ring structure.

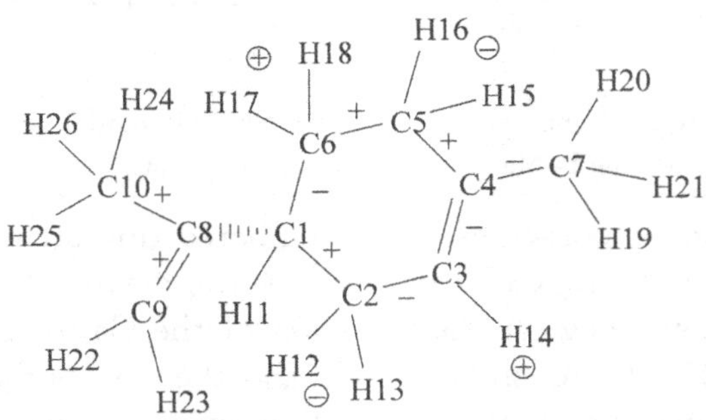

Fig. 8. 4 The signs of the differential bond polarizabilities, (+, −) for C—C stretching coordinates and (⊕, ⊖) for the bending coordinates (C6—H_2 bending, C2—H_2 bending, C5—H_2 bending, C3—H bending).

Comments

(1) The show-up of intramolecular enantiomerism is rather unexpected since the breaking of the mirror symmetry by the structural replacements/substituents is in no sense a small perturbation. The perturbation is just not strong enough to break the sign opposition of the differential bond polarizabilities of the otherwise mirror symmetry re-

lated pair bond coordinates. The small residual differences between the pair differential bond polarizabilities are of course responsible for the generation of the ROA signals.

(2) We have to acknowledge that the ROA spectrum of (R)-limonene implies the existence of an approximate mirror symmetry which is not exact due to the replacements/substituents on its ring. This means that ROA spectrum does contain stereo-structural information.

(3) We have tried the analysis by the Gaussian09W calculated Raman and ROA intensities. The results are shown in Fig. 8. 5 and Fig. 8. 6. For the C—H stretching part, the enantiomerism is evident. However, for the ring C—C stretching and the C—H bending parts, there is variation. This demonstrates the delicate variation between the experimental and calculated intensities.

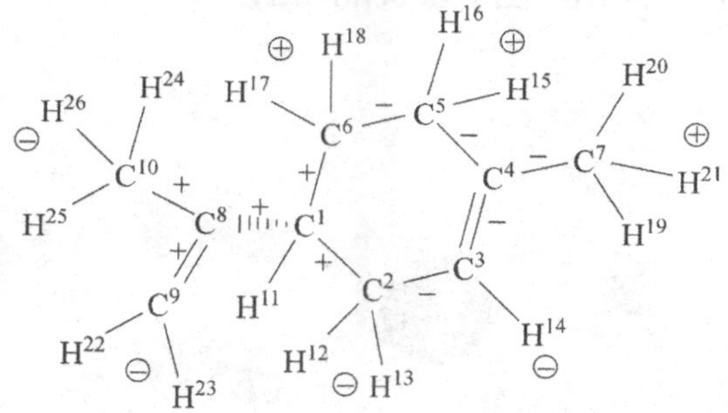

Fig. 8. 5 The signs of the differential bond polarizabilities. (+, −) for the C—C stretching, (⊕,⊖) for the C—H bendings.

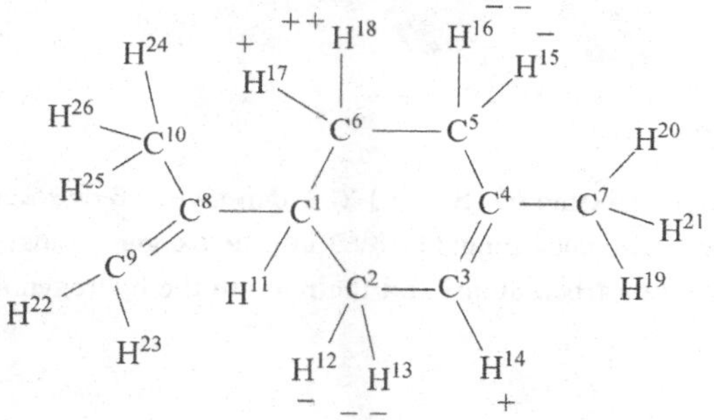

Fig. 8. 6 The signs of the differential bond polarizabilities of the C—H stretchings. "+", "−" are for the $C—H_2$ symmetric stretching and "++", "− −" are for the $C—H_2$ anti-symmetric stretching.

8.3 The case of (S)-(+)-2, 2- dimethyl -1, 3- dioxolane-4-methanol

The structure of (S)-(+)-2, 2-dimethyl-1, 3-dioxolane-4-methanol was optimized by DFT B3LYP at the level of 6-31g+(d, p) by Gaussian09W. Gaussian09W was also employed to simulate its Raman and ROA wavenumbers and intensities. Its structure and atomic numberings are shown in Fig. 8.7. Listed in Table 8.3 are the Raman, ROA wavenumbers, intensities and PED (potential energy distribution) of the modes which were employed in our analysis. Shown in Table 8.4 are the symmetrized coordinates which were used in the normal mode analysis. From Table 8.3, we note that the C—H and O—H stretchings are decoupled from the skeletal C—C and C—O stretchings. Hence, they will be treated independently.

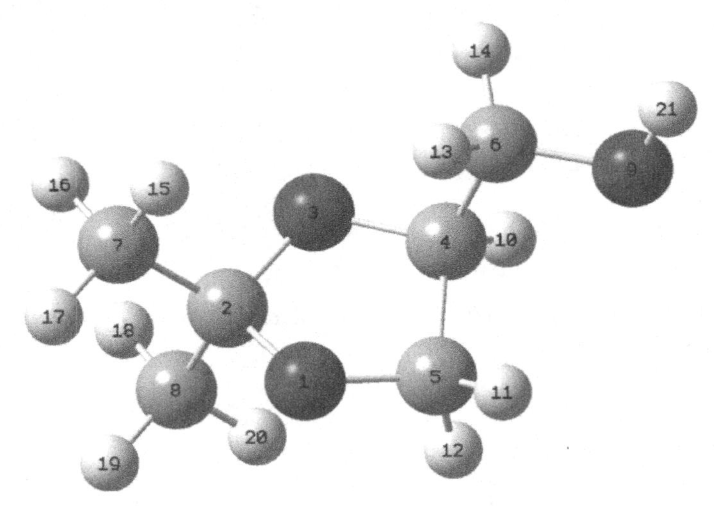

Fig. 8.7 The structure of (S)-(+)-2, 2-dimethyl-1, 3-dioxolane-4-methanol and its atomic numberings. 1, 3, 9 are the oxygen atoms; 2, 4, 5, 6, 7, 8 are the carbon atoms and the rest are the hydrogen atoms.

Table 8.3 The calculated Raman, ROA wavenumbers, intensities and PED(potential energy distribution) of the modes of (S)-(+)-2, 2-dimethyl-1, 3-dioxolane-4-methanol which are employed in the analysis. Only those PED that are larger than 10 are listed.

Raman wavenumber	Raman intensity	ROA intensity	PED
3847.4	618.2	389.2	S_{21}(100.0)
3134.5	392.1	643.9	S_{11}(25)S_{12}(68)
3071.6	409.1	169.8	S_{10}(68)S_{14}(29)
3059.4	1078.5	67.6	S_{15}(84)S_{18}(14)
3053.9	475.0	19.9	S_{10}(26)S_{14}(66)
3051.1	184.9	58.7	S_{15}(13)S_{18}(81)
3025.2	371.6	1251.3	S_{11}(69)S_{12}(28)
3011.9	446.9	−265.9	S_{13}(97)
1277.6	30.1	91.6	S_7(25)S_{27}(11)S_{52}(13)
1232.7	35.6	−194.1	S_8(27)S_{27}(15)S_{46}(21)
1170.4	1.0	−123.9	S_{28}(19)S_{47}(24)S_{51}(14)
1095.1	9.3	70.9	S_4(13)S_5(44)S_{23}(14)
1072.8	6.5	97.6	S_2(10)S_3(58)S_{22}(10)S_{40}(12)
1058.7	38.4	−157.9	S_9(100)
1030.1	14.5	112.2	S_4(14)S_5(32)S_6(14)
968.8	32.1	−128.6	S_7(14)S_8(10)S_{52}(20)
950.4	3.8	−15.6	S_{35}(36)S_{37}(15)S_{40}(25)
924.8	25.0	3.3	S_7(12)S_8(20)S_{46}(39)S_{52}(23)
850.7	17.5	−159.5	S_1(27)S_2(35)S_4(11)S_6(10)S_{47}(10)
848.5	30.9	15.2	S_1(13)S_4(34)S_6(13)
793.7	34.7	8.3	S_1(20)S_2(21)S_{23}(17)
750.1	7.6	−50.6	S_{22}(17)S_{23}(37)S_{40}(10)
650.3	29.9	−3.3	S_7(19)S_8(12)S_{22}(34)
507.5	3.8	−10.7	S_{27}(77)

Table 8.4 The locally symmetrized coordinates of (S)-(+)-2, 2-dimethyl-1, 3-dioxolane-4-methanol. r and α are the bond stretching and bending coordinates. $a=\cos 144°$, $b=\cos 72°$.

Definition	Description
$S_1=r(C2-O1)$	C2—O1 stretching
$S_2=r(C2-O3)$	C2—O3 stretching
$S_3=r(C4-O3)$	C4—O3 S, retching
$S_4=r(C4-C5)$	C4—C5 stretching
$S_5=r(C5-O1)$	C5—O1 stretching
$S_6=r(C4-C6)$	C4—C6 stretching
$S_7=r(C2-C7)$	C2—C7 stretching
$S_8=r(C2-C8)$	C2—C8 stretching
$S_9=r(C6-O9)$	C6—O9 stretching
$S_{10}=r(C4-H10)$	C4—H10 stretching
$S_{11}=r(C5-H11)+r(C5-H12)$	C5—H_2 symmetric stretching
$S_{12}=r(C5-H11)-r(C5-H12)$	C5—H_2 antisymmetric stretching
$S_{13}=r(C6-H13)+r(C6-H14)$	C6—H_2 symmetric stretching
$S_{14}=r(C6-H13)-r(C6-H14)$	C6—H_2 antisymmetric stretching
$S_{15}=r(C7-H15)+r(C7-H16)+r(C7-H17)$	C7—H_3 symmetric stretching
$S_{18}=r(C8-H18)+r(C8-H19)+r(C8-H20)$	C8—H_3 symmetric stretching
$S_{21}=r(O9-H21)$	O9—H21 stretching
$S_{22}=\alpha(C5O1C2)+a[\alpha(O1C2O3)+\alpha(C4C5O1)]$ $+b[\alpha(C2O3C4)+\alpha(O3C4C5)]$	ring deformation
$S_{23}=(a-b)[\alpha(O1C2O3)-\alpha(C4C5O1)]$ $+(1-a)[\alpha(C2O3C4)-\alpha(O3C4C5)]$	ring deformation
$S_{27}=\alpha(O1C2C7)-\alpha(O1C2C8)$ $+\alpha(O3C2C7)-\alpha(O3C2C8)$	C2—C7—C8 rocking
$S_{40}=\alpha(C4C6H13)-\alpha(C4C6H14)$ $+\alpha(O9C6H13)-\alpha(O9C6H14)$	C6—H_2 rocking
$S_{46}=2\alpha(C2C7H15)-\alpha(C2C7H16)-\alpha(C2C7H17)$	C7—H_3 rocking
$S_{47}=\alpha(C2C7H16)-\alpha(C2C7H17)$	C7—H_3 rocking
$S_{52}=\alpha(C2C8H19)-\alpha(C2C8H20)$	C8-H_3 rocking

For the C—H and O—H portions, we have eight stretching coordinates: C4—H10, C5—H11, C5—H12, C6—H13, C6—H14, C7—H_3 (symmetric), C8—H_3 (symmetric), O—H to solve for their bond polarizabilities. The criterion for solving the phase problem is that the elucidated bond polarizabilities of C—H and O—H are positive. We have three phase sets left. Their bond polarizabilities and differential bond polarizabilities are listed in Table 8.5(a). Considering that C6—H13 and C6—H14 stretchings; and those of C7—H_3 (symmetric stretching) and C8—H_3 (symmetric stretching) are quite equivalent, their bond polarizabilities should not differ too much, Solution 3 is the most possible one.

Table 8.5 (a) The bond polarizabilities of C—H, O—H (the 1st row) and their differential bond polarizabilities (the 2nd row). (b) The bond polarizabilities of the skeletal bonds and bendings (the 1st and 2nd rows) and their differential bond polarizabilities (the 3rd and 4th rows). The bond polarizabilities and differential bond polarizabilities are relative.

(a)

Solution 1							
C4—H10	C5—H11	C5—H12	C6—H13	C6—H14	C7—H_3	C8—H_3	O9—H21
28.03	546.24	309.15	5.05	799.97	355.35	254.88	671.41
C4—H10	S11	S12	S13	S14	S15	S18	O9—H21
−1.60	11.29	−0.79	−1.17	−1.35	0.12	−0.23	4.24
Solution 2							
594.78	571.91	404.43	484.09	95.75	416.04	35.88	671.80
−1.11	11.25	−0.77	−1.25	−0.93	0.49	−1.18	4.24
Solution 3							
503.72	584.28	438.64	435.48	169.06	288.65	348.44	671.79
−1.40	11.32	−0.80	−1.21	−1.12	0.09	−0.19	4.24

(b)

Solution 1								
C2—O1	C2—O3	C4—O3	C4—C5	C5—O1	C4—C6	C2—C7	C2—C8	C6—O9
S22	S23	S27	S40	S46	S47	S52		
73.20	203.90	66.68	121.60	248.50	88.37	61.87	53.45	183.42

−5.51	60.39	67.83	34.54	−38.42	−17.75	−16.67		
−15.09	9.21	16.73	4.78	−13.59	−14.59	4.16	−6.80	−9.72
2.12	−0.97	−0.51	2.22	2.40	9.55	−2.91		
Solution 2								
49.91	229.44	68.29	98.82	234.59	64.07	63.15	49.59	179.27
−6.10	59.65	68.01	36.29	−39.01	−26.28	−16.28		
−6.21	−0.53	16.12	13.47	−8.28	−5.33	3.67	−5.32	−8.13
2.35	−0.69	−0.57	1.55	2.63	12.80	−3.06		
Solution 3								
111.00	233.08	120.69	1.25	130.33	245.32	69.27	56.53	213.67
−12.32	66.91	62.52	23.75	−39.55	−22.74	−14.99		
−12.17	11.46	20.91	−4.53	−22.73	−2.45	4.73	−6.56	−7.38
1.60	−0.47	−0.92	1.39	2.32	9.17	−2.78		
Solution 4								
91.64	247.86	141.47	0.13	114.54	244.34	74.85	52.01	212.95
−10.54	65.86	62.04	23.16	−37.93	−2.68	−15.77		
11.82	−6.85	−4.84	−3.15	−3.16	−1.23	−2.18	−0.95	−6.49
−0.61	0.84	−0.33	2.12	0.31	−15.70	−1.81		

For the skeletal C—C and C—O portions, we have nine stretching coordinates: C2—O1, C2—O3, C4—O3, C4—C5, C5—O1, C4—C6, C2—C7, C2—C8 and C6—O9 to solve. From Table 8.3, we note that S22, S23, S27, S40, S46, S47 and S52 bending coordinates are significantly coupled to these C—C and C—O stretchings. Totally, we have 16 coordinates to solve for their bond polarizabilities. The criteria for solving the phases are that the bond polarizabilities of C—C and C—O are positive and the differences among the C—O stretchings are no more than 5 fold. Meanwhile, we limit the difference between those of C2—C7, C2—C8 stretchings to be smaller than 2 fold since they are quite equivalent. (It should be stressed that these figures are not crucial.) We further require that the signs of those of S47 (C7—H_3 rocking) and S52 (C8—H_3 rocking) are the same since they are of the same bending type. By this way, we are left with 4 phase sets. They are listed in Table 8.5(b). We note that in Solution 3, the bond polarizability of C4—C5 is too small and in Solution 4, those of S47 and S52 are too impartial. As to the left Solutions 1 and 2, Solution 1 is the most adequate one since its bond polarizabilities for S47 and S52 are most close to each other.

We will pay attention to the signs of the differential bond polarizabilities. They are shown in Fig. 8. 8. Evidently, the signs for the two pair C—O bonds(in the figure, the upper and lower C—O bonds) are opposite. The corresponding mirror for this sign inversion is the one connecting the asymmetric atom(C2) and the midpoint of the C4—C5 bond and vertical to the (approximate) ring structural plane. Moreover, the sign for the C5—H_2 symmetric stretching is positive. It is just opposite to the sign for the C4—H10 stretching coordinate. The two C—H_3 groups are above and below the roughly planar ring plane. The differential bond polarizability signs of their symmetric stretching and bending coordinates are opposite. The signs for the two C—C bond coordinates that connect these two methyl groups to the asymmetric atom are also opposite. All these show the same phenomenon as happens in the limonene case, i. e. , the intramolecular enantiomerism is confirmed here.

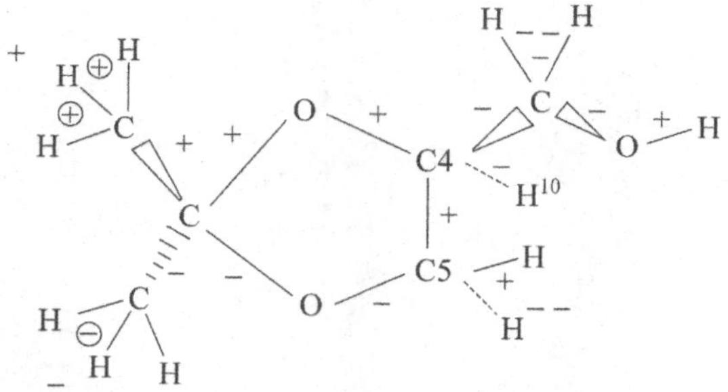

Fig. 8. 8 The signs of the differential bond polarizabilities. (+), (−) for C—C, C—O, O—H bond stretchings and C—H_3, C—H_2 symmetric stretching coordinates. (− −) for the C—H_2 antisymmetric stretching coordinate. (⊕), (⊖) for the C—H_3 bending coordinates.

The magnitude difference between the differential bond polarizabilities of the pair coordinates related by the intramolecular mirror reflection is an indication of the degree of Raman optical activity as far as bond coordinates are considered. It shows the core of ROA phenomenon. For our case of 2, 2- dimethyl- 1, 3- dioxolane -4- methanol, this is shown in Fig. 8. 9(a). From it, we note that as the pair bond coordinates are farther away from the asymmetric atom, this magnitude difference is smaller. Shown in Fig. 8. 9(b) is the case of (R)-limonene. The same inference is confirmed. The central role played by the asymmetric atom(center) in ROA is confirmed hereby. We stress that this is a vivid spectroscopic demonstration of the asymmetry of the chiral center.

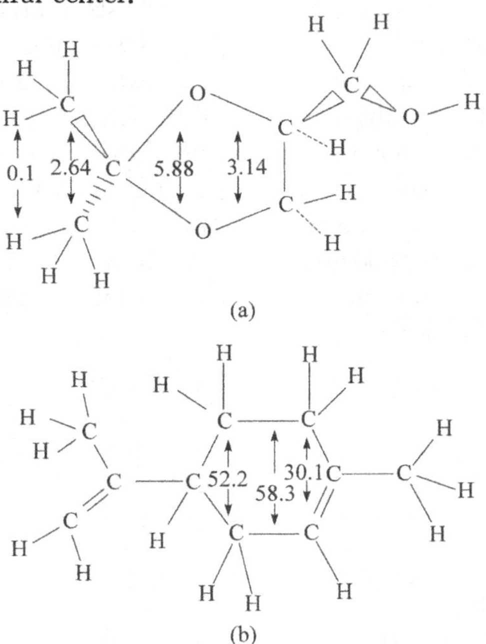

Fig. 8. 9 The magnitude differences between the differential bond polarizabilities of the pair coordinates related by the intramolecular mirror reflection. (a) 2, 2- dimethyl- 1, 3- dioxolane -4- methanol, (b) (R)-limonene.

Comments

In this case study of 2, 2- dimethyl- 1, 3- dioxolane -4- methanol, besides the confirmation of intramolecular enantiomerism, the central ROA role played by the asymmetric atom(center)is confirmed from a spectroscopic viewpoint: as the pair bond coordinates, that are mirror reflected to each other, are farther away from the asymmetric atom, their differential bond polarizability difference is smaller. We note that, in general, this asymmetry is but mentioned from a geometric viewpoint: the non-superimposability of the mirror images around the asymmetric atom, or equivalently, that the four bonds connected to the asymmetric carbon atom are different.

8.4 More cases for intramolecular enantiomerism

We just show our results for the two cases of (R)-(+)-4-isopropyl-1-methyl cyclohexene and (R)-(+)-3-methylcyclohexanone. Their Raman and ROA intensities are simulated by Gaussian09W software.

Shown in Fig. 8. 10 are the structure and signs of the differential bond polarizabilities of(R)-(+)-4-isopropyl-1-methylcyclohexene. We note that its ring structure is not planar and the mirror along C1 and C4(or along the mid-points of C5—C6 and C2—C3 in Fig. (c))is not exact. However, the intramolecular enantiomerism is very evident, even for the two methyl groups attached to C8 which are away from the ring moiety.

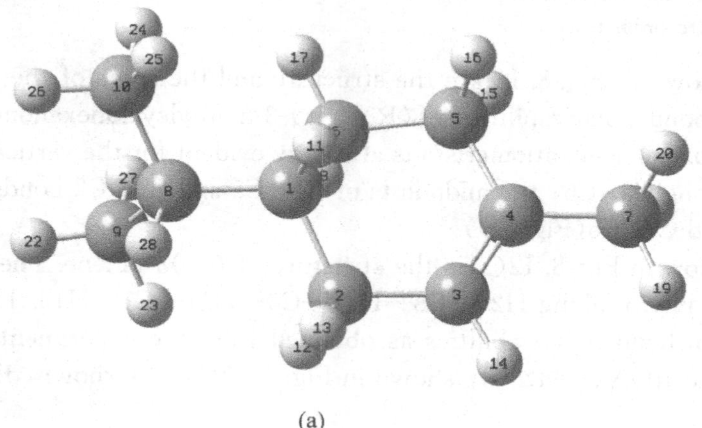

(a)

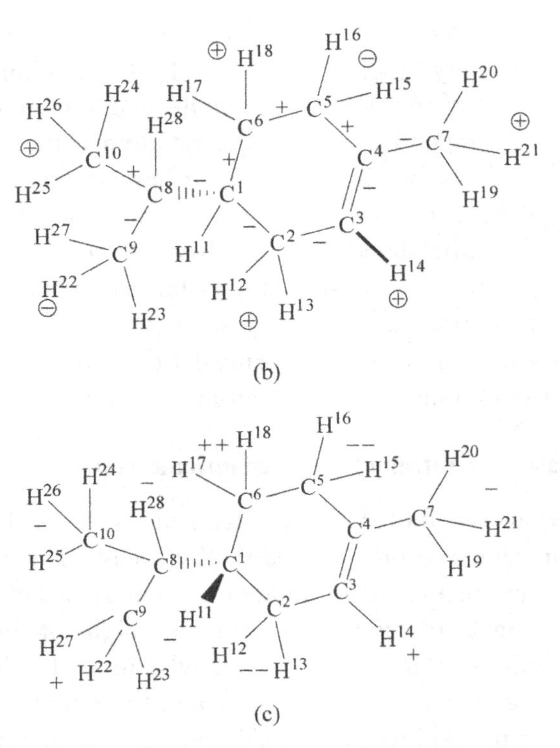

Fig. 8. 10 The structure of (R)-(+)-4-isopropyl-1-methylcyclohexene (a) and the signs of its differential bond polarizabilities, +/− are for the bond stretching, ⊕/⊖ are for the C—H, C—H_2 and C—H_3 bendings(b). ++/− − are for the C—H_2 anti-symmetric coordinate and +/− are for the C—H, C—H_3 symmetric (stretch) coordinates(c).

Shown in Fig. 8. 11 are the structure and the signs of the differential bond polarizabilities of (R)-(+)-3-methylcyclohexanone. The intramolecular enantiomerism is also very evident for the vertical mirrors along C3, C6; the midpoints of C4—C5 and C1—C2 bonds(Fig. (b)) and C2, C5(Fig. (c)).

Show in Fig. 8. 12(a) is the structure of (−)α-pinene. There is a pseudo mirror along H20—C8—H18—C6—H17—C3—H13. The differential bond polarizabilities as obtained from the experimental Raman and ROA spectra are shown in Fig. 8. 12(b). As shown, there is

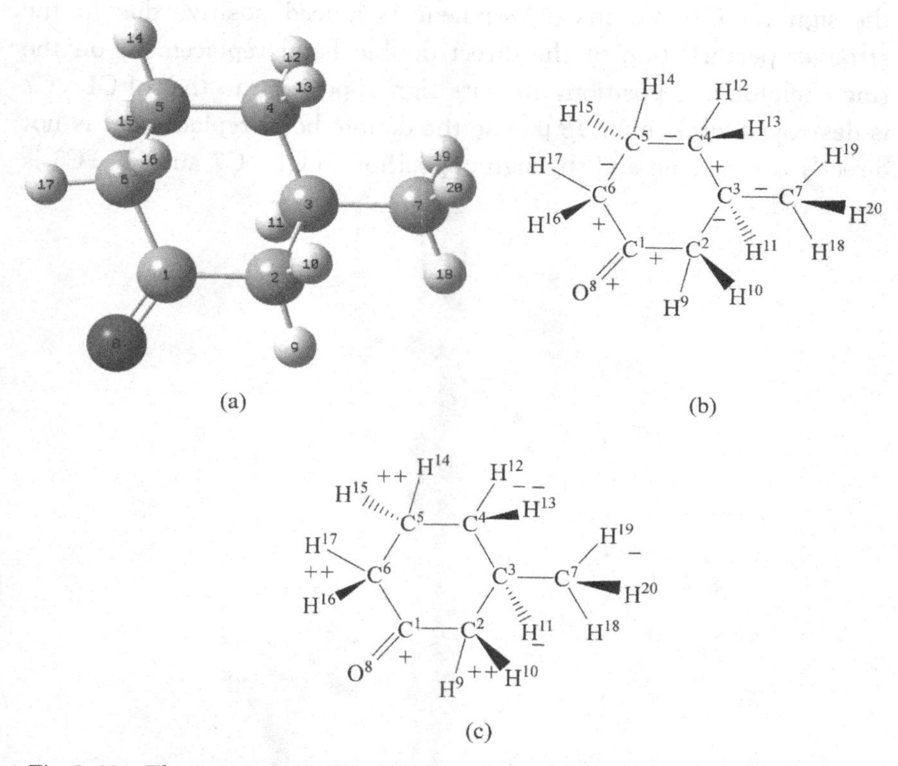

Fig. 8. 11 The structure of (R)-(+)-3-methylcyclohexanone(a) and the signs of its differential bond polarizabilities, +/− are for the C—C, C—O bond stretching coordinates(b). ++/− − are for the C—H_2 anti-symmetric coordinate, −is for the C3—H stretching and C7—H_3 symmetric(stretch) coordinates, + is for the C—O stretching coordinate(c).

evidence of intramolecular enantiomerism in (−) α-pinene, for the pair bonds: C3—C4, C1—C3; C5—C6, C6—C7; C5—C8, C7—C8. In fact, the differential bond polarizability of C4—C5, though positive, is very close to 0, so its sign opposition to that of C1—C7 which is positive, cannot be absolutely denied. Those of (−) β pinene, obtained also from the experimental Raman and ROA spectra are shown in Fig. 8. 12(c). Its intra-molecular enantiomerism is also very evident. The comparison of these two cases may lead to the inference that the sign for C4—C5 in (−) α-pinene is indeed positive due to the stronger perturbation of the direct double bond replacement on the ring skeleton at α position, thus its sign opposition to that of C1—C7 is destroyed while in(−)β pinene the double bond replacement is not directly on the ring and the sign opposition in C1—C7 and C4—C5 is preserved.

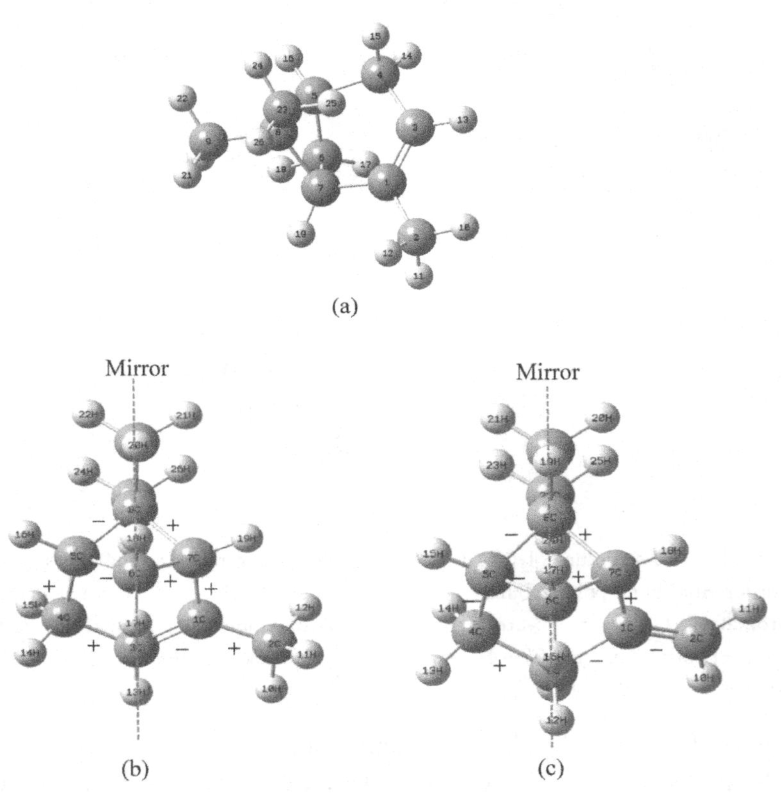

Fig. 8. 12 The structure of (−)α-pinene of which the atoms labeled from 1 to 9 and 23 are carbon(C) atom and the rest are hydrogen(H) atom(a), its signs of differential bond polarizabilities(b) and those of(−) β-pinene(c)

Comments

(1) We appreciate that through the bond polarizability analysis we can retrieve the physical picture behind the Raman/ROA intensities. The demonstration of intramolecular enantiomerism is not trivial.

(2) For a molecule possessing a mirror symmetry, its vibrationally induced electric and magnetic couplings for ROA mechanism, say, on the two sides of the mirror are just of opposite signs but the same magnitude. Their cancellation thus causes achirality. As this mirror symmetry is broken, the cancellation is not complete and chirality or ROA appears. However, this symmetry breaking may not be so serious that the sign opposition of the electric and magnetic couplings on the two sides of the now non-exact mirror is still preserved. This sign opposition is reflected in the differential bond polarizabilities. Of course, if the symmetry breaking is strong enough, this phenomenon will disappear, i. e. , the signs of the differential bond polarizabilities will be in a random way. Our results show that in general the symmetry breaking is just not so strong and intramolecular enantiomerism can be general.

Chapter 9

A unified classical theory for ROA and VCD

9.1 Background

The difference between ROA and VCD (vibrational circular dichroism) is that in ROA, there is charge excitation/disturbance while in VCD, it is the charges in the ground state that is involved in the chiral infrared absorption. The charges involved in VCD can be attributed to the Mulliken charges in the ground state from a conceptual/classical picture. We know that bond polarizability is a measure of charge excitation during the Raman process. We may approximate charge re-distribution in Raman by partitioning the bond polarizabilities after proper scaling to the atoms, and then combining them with the Mulliken charges on the atoms in the ground state. Though this is a conceptual or classical picture, we will find that based on it, one can develop a classical algorithm that can offer a simple but clear physical picture for both ROA and VCD mechanisms on equal footing.

For ROA, the following parameter is commonly defined for its characterization:

$$\Delta \equiv (I^R - I^L)/(I^R + I^L).$$

Here, I^L and I^R are the Raman mode intensities under left and right circularly polarized laser excitations, respectively. Of course, various geometric arrangements of ROA experimental setup can lead to different definitions for Δ and will reveal various aspects and properties of ROA. Since we only concern the classical interpretation for ROA, these details will not be over-emphasized. As our goal is not on the quantitative accuracy of Δ in comparison with the experimental observation, we will mainly emphasize the physical picture of ROA and compare the Δ sign by our classical algorithm with that from the experimental observation. We will use Δ_{cl} to denote the Δ sign by our classical algorithm. For VCD, analogous Δ parameter can be defined by substituting the absorption intensity for the Raman intensity.

Comments

The question why we are stuck on a classical algorithm instead of modern quantum mechanical calculation may be raised. For this, we have to confess that in molecular physics, though many physical properties can be deduced based on quantal calculation, but not all are necessarily to be done in this way. Indeed, many can still be deduced in a classical way. The merit of adopting the classical way is that often it can offer a comprehensive physical picture for the topic that is of our concern.

9.2 The classical algorithm

The physical picture of ROA mechanism is shown below. The j-th atom in a molecule with net charge e_j, under a normal mode vibration, will oscillate around its equilibrium position periodically. Its displacement $\Delta \boldsymbol{r}_j$ and the induced dipole $\boldsymbol{\mu}_j$ (i. e. , $e_j \Delta \boldsymbol{r}_j$) are coincident in space. Meanwhile, the other i-th atom, with net charge e_i, will produce a magnetic moment $\boldsymbol{m}_{ij}$ on the j-atom, with a value of

$$\frac{1}{2}(e_i \boldsymbol{v}_i \times \boldsymbol{r}_{ij}).$$

Here, $\boldsymbol{r}_{ij}$ is the position vector from the i-th to the j-th atoms. $\boldsymbol{v}_i$ is the velocity of the i-th atom. The quantal expression for the intra-molecular interaction between the vibrationally induced electric and magnetic dipoles that is related to the ROA mechanism is[1]:

$$\mathrm{Im}(\langle i | \boldsymbol{\mu} | f \rangle \langle f | \boldsymbol{m} | i \rangle).$$

Here, $|i\rangle$ and $|f\rangle$ are the initial and final states. Im shows the imaginary part of the electromagnetic interaction, for which $\boldsymbol{\mu}$ and $\boldsymbol{m}$ are out of phase by $\pi/2$. Hence, its classical expression is

$$\sum \boldsymbol{\mu} \cdot \boldsymbol{m}.$$

Specifically, for a normal mode, the term that determines its Δ_d sign is:

$$\sum_j \sum_i \boldsymbol{\mu}_{j\max} \cdot \boldsymbol{m}_{ij\,\mathrm{equ}}.$$

Here, $\boldsymbol{\mu}_{j\max}$ is the dipole when $\Delta \boldsymbol{r}_j$ reaches its maximum (denoted as $\Delta \boldsymbol{R}_j$). $\boldsymbol{m}_{ij\,\mathrm{equ}}$ is the magnetic moment induced by the motion of the charge on the i-th atom at the j-th atom when $\Delta \boldsymbol{r}_i = 0$ with maximal $\boldsymbol{v}_i$. (Note the phase difference of $\pi/2$ between $\boldsymbol{\mu}_{j\max}$ and $\boldsymbol{m}_{ij\,\mathrm{equ}}$.) When $\Delta \boldsymbol{r}_i = 0$, $\boldsymbol{v}_i$ is proportional to $\Delta \boldsymbol{R}_i$. (As $\Delta \boldsymbol{r}_i = 0$, $\boldsymbol{v}_i$ is proportional to $\Delta \boldsymbol{R}_i / \tau$. τ is the period of the normal mode.) Hence, Δ_d is determined by the sign of the following expression:

$$\begin{aligned} &\sum_{ji} (e_j \Delta \boldsymbol{R}_j) \cdot (e_i \Delta \boldsymbol{R}_i \times \boldsymbol{r}_{ij}) \\ = &\sum_{ji} e_i e_j \boldsymbol{r}_{ij} \cdot (\Delta \boldsymbol{R}_j \times \Delta \boldsymbol{R}_i). \end{aligned}$$

Since $|\Delta \boldsymbol{R}_i|$ ($|\Delta \boldsymbol{R}_j|$) is much smaller than $|\boldsymbol{r}_{ij}|$, $\boldsymbol{r}_{ij}$ can take the value at the molecular equilibrium configuration. In principle, $\boldsymbol{r}_{ij}$ can be obtained from the molecular configuration or by the quantum chemical calculation. $\Delta \boldsymbol{R}_i$ ($\Delta \boldsymbol{R}_j$) can be obtained from the normal mode analysis, if force constants are known. As to the chirality of the above formula, we note that $\boldsymbol{r}_{ij} \cdot (\Delta \boldsymbol{R}_j \times \Delta \boldsymbol{R}_i)$ changes its sign as the coordinate system is shifted between the right-hand one and the left one as anticipated. We will see later that for ROA and VCD, the interpretations for e_i are different. In the following section, we will first consider the situation due to the charge re-distribution in the Raman excitation for ROA.

Comments

The above classical treatment is limited to the coupling between the vibrationally induced electric and magnetic dipole moments, excluding higher couplings involving the quadrupole moment. Besides, only the net charges on the atoms are considered. No consideration is paid to the effect by the charges distributing spatially on the bonds and the magnetic effect due to the intra-molecular electric current which is caused by the charge redistribution when the molecule vibrates, if there is any[2, 3].

9.3 The charge re-distribution in the Raman excited virtual state

In our proposed classical ROA algorithm, we only consider the charges on each atom. Under this approximation, for considering the charge re-distribution in the virtual state as evidenced by the bond polarizabilities, we will first partition each bond polarizability (since charges are concentrated on the bonds, we will only employ the bond polarizabilities of the stretching coordinates in this treatment) evenly to the two atoms forming the bond. Then, all the polarizabilities on an atom are summed, scaled and combined with its net charges in the ground state. (Since bond polarizabilities obtained are on a relative scale, they need to be scaled if they are to be compared to the electron numbers in the molecule. We will show how this scaling can be nailed down.)

For the moment, we pay more attention to the charges on the atoms. When a molecule is in its ground state, there are charges distributing in between the atoms. This is the idea of the chemical bond. For convenience, one usually attributes the charges on a bond to the atoms which form the bond. This is the Mulliken charge(e_j^0, subscript j is for the atomic numbering). So, it is an effective charge. Since a molecule is neutral, the sum of all the Mulliken charges is zero, i. e., $\sum_j e_j^0 = 0$. In the Raman virtual state, the sum of the net charges on all the atoms is still zero. This means that the charges on the atoms which are derived from the bond polarizabilities after scaling as shown previously have to come from the Mulliken charges in the ground state. We know that close to the complete relaxation of the Raman excitation, the bond polarizabilities are very parallel to the bond electronic densities in the ground state which are calculated by all the occupied MO, i. e., all the electrons in a molecule. (At least, in the case we will deal with.) This implication is that all the electrons in a molecule contribute to the excited charges in the Raman virtual state with equal probability. The electron number on the j-th atom in the ground state is $z_j = Z_j - e_j^0$ with Z_j the atomic number. Hence, in the virtual state, the excited charges contributed from the j-th atom will be

$$[t \cdot z_j \cdot \sum_l \partial\alpha/\partial R_l]/N \quad (t > 0)$$

where N is the total electron number in a molecule, i. e., $N = \sum_j Z_j (= \sum_j z_j)$, $\sum_l \partial\alpha/\partial R_l$ is the sum of all the bond polarizabilities and t is the scaling factor. Since the distributions of the excited charges on the various bonds (as evidenced by the bond polarizabilities) are different and if we partition each bond polarizability evenly to the two atoms forming the bond, then the effective charge on the j-th atom is

$$e_j = e_j^0 + [t \cdot z_j \cdot \sum_l \partial\alpha/\partial R_l]/N - t\sum_{k_j} (\partial\alpha/\partial R_{k_j})/2$$

where k_j is the index for the bonds that are associated with the j-th atom. We note that

$$\sum_j z_j \cdot \{[\sum_l \partial\alpha/\partial R_l]/N\} = \sum_j \sum_{k_j} (\partial\alpha/\partial R_{k_j})/2 = \sum_l \partial\alpha/\partial R_l,$$

and $\sum_j e_j^0 = 0$, hence $\sum_j e_j = 0$. This satisfies the condition that the net charge of the Raman virtual state is zero. We note that if $z_j \cdot [\sum_l \partial\alpha/\partial R_l]/N - \sum_{k_j} (\partial\alpha/\partial R_{k_j})/2$ is negative, then the j-th atom will acquire electrons in the Raman excitation, otherwise, it will lose electrons. As to t, if we scale the bond polarizabilities to the unit of the elementary charge, then we have $t \cdot \sum_l \partial\alpha/\partial R_l \leqslant N$, or $t \leqslant N/\sum_l \partial\alpha/\partial R_l$.

Evidently, in VCD, only $e_j = e_j^0$ will be needed for the Δ_d calculation by the formula developed in the previous section. In this aspect, we note that our approach puts ROA and VCD mechanisms on equal footing.

Comments

We may appreciate that our treatment puts ROA and VCD mechanisms on equal footing. Though this is classical, however, it does offer us a very clear physical picture for both ROA and VCD.

9.4 The ROA application on (+)-(R)-methyloxirane

The molecular system we will try on is(+)-(R)-methyloxirane whose structure together with its atomic numberings is shown in Fig. 9.1. We have recorded both its Raman and ROA spectra by the 532 nm excitation. From its Raman(relative) intensities, the(relative) bond polarizabilities were determined. The relative bond polarizabilities with that of C2O1 normalized to 100 are listed in Table 9.1. (Refer to Section 6.3) The bond polarizabilities, as interpreted in the previous section, reflect the re-distribution of the excited charges in the Raman excitation. As demonstrated in the previous section, the effective charge e_j on the j-th atom in the Raman excitation is

$$e_j = e_j^0 + [t \cdot z_j \cdot \sum_l \partial\alpha/\partial R_l]/N - t\sum_{k_j}(\partial\alpha/\partial R_{k_j})/2$$

The Mulliken charges (e_j^0) were calculated by the DFT quantum method based on 6-31G* basis. For this molecule, $\sum_l \partial\alpha/\partial R_l = 459.64$ as the bond polarizability of C2O1 is normalized to 100. $N=32$ ($N=6\cdot 3+8\cdot 1+1\cdot 6=32$) and t has to be less than 0.0696 (Note that $t \leqslant N/\sum_l \partial\alpha/\partial R_l$). Other parameters were also calculated and listed in Table 9.1. The calculation of $z_j \cdot [\sum_l \partial\alpha/\partial R_l]/N - \sum_{k_j}(\partial\alpha/\partial R_{k_j})/2$ shows that in the Raman excitation, C2, C3, H4, H5, H6, especially, C2 and H6 acquire electrons while the rest atoms, especially, C7 and O1 lose electrons.

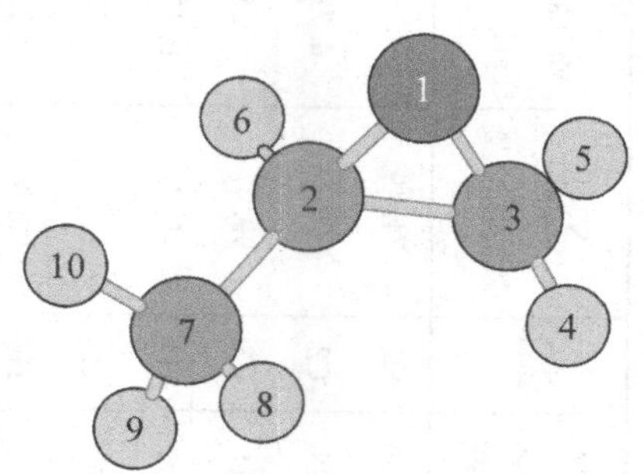

Fig. 9.1 The structure of(+)-(R)-methyloxirane and its atomic numberings. Atom 1 is oxygen, atoms 2, 3, 7 are carbon and the rest are H atoms.

Table 9. 1　For (+)-(R)-methyloxirane, the relative bond polarizabilities, the polarizability attributed to each atom ($\sum_{k_j}(\partial\alpha/\partial R_{k_j})/2$), the Mulliken charge($e_j^0$), the electron number($z_j = Z_j - e_j^0$ with Z_j the atomic number) on each atom based on the Mulliken charge, the redistribution of the charges as evidenced by the bond polarizabilities in the virtual state ($z_j \cdot [\sum_l \partial\alpha/\partial R_l]/N - \sum_{k_j}(\partial\alpha/\partial R_{k_j})/2$, where $\sum_l \partial\alpha/\partial R_l = 459.64$ is the sum of all the bond polarizabilities.) and the effective charge($e_j = e_j^0 + [t \cdot z_j \cdot \sum_l \partial\alpha/\partial R_l]/N - t\sum_{k_j}(\partial\alpha/\partial R_{k_j})/2$) range as the scaling factor t varies from 0. 0133 to 0. 0160.

Bond	Relative bond polarizability	Atom	$\sum_{k_j}(\partial\alpha/\partial R_{k_j})/2$	Mulliken charge(e_j^0)	z_j	$z_j \cdot [\sum_l \partial\alpha/\partial R_l]/N - \sum_{k_j}(\partial\alpha/\partial R_{k_j})/2$	$e_j = e_j^0 + [t \cdot z_j \cdot \sum_l \partial\alpha/\partial R_l]/N - t\sum_{k_j}(\partial\alpha/\partial R_{k_j})/2$ with t=0. 0133 and 0. 0160	
C2O1	100. 00	O1	74. 49	−0. 84	8. 84	52. 49	−0. 15	−0. 01
C3O1	48. 98	C2	144. 46	0. 72	5. 28	−68. 62	−0. 18	−0. 37
C2C3	87. 11	C3	99. 47	−0. 14	6. 14	−11. 28	−0. 30	−0. 33
C2H6	79. 80	H4	15. 71	0. 15	0. 85	−3. 50	0. 11	0. 10
C2C7	22. 00	H5	15. 71	0. 16	0. 84	−3. 65	0. 12	0. 11
C3H4	31. 43	H6	39. 90	0. 16	0. 84	−27. 84	−0. 20	−0. 28
C3H5	31. 43	C7	40. 45	−0. 71	6. 71	55. 93	0. 03	0. 18
C7H8	19. 63	H8	9. 82	0. 16	0. 84	2. 25	0. 19	0. 20
C7H9	19. 63	H9	9. 82	0. 16	0. 84	2. 25	0. 20	0. 20
C7H10	19. 63	H10	9. 82	0. 16	0. 84	2. 25	0. 20	0. 21

Then the Δ_d sign of a normal mode can be readily determined by the sign of $\sum_{ji} e_i e_j \boldsymbol{r}_{ij} \cdot (\Delta \boldsymbol{R}_j \times \Delta \boldsymbol{R}_i)$ which can be calculated as soon as the normal mode analysis for this molecule is done, with a presumed t factor. In the calculation for Δ_d, we found that if the scaling factor t is in the range of(0. 0133, 0. 0160)(in calculation, the digital step is 0. 0001), then the experimental ROA signs can be well reproduced by the calculated Δ_d. This is indeed remarkable as shown in Table 9. 2! Shown in Table 9. 2 are also the calculated Δ_d signs simply by the Mulliken charges without those originating from the bond polarizabilities. Then, the discrepancy with the experimental observation is obvious. This indicates that in the Raman excitation, the charge re-distribution is not negligible. The charge re-distribution is evidenced by the bond polarizabilities. These charge modifications on the atoms in the Raman excited state are significant. They are of the same order as the Mulliken charges as shown in Table 9. 1 by the comparison of e_j^0 and $[t \cdot z_j \cdot \sum_l \partial\alpha/\partial R_l]/N - t\sum_{k_j} (\partial\alpha/\partial R_{k_j})/2$.

Table 9. 2 For(+)-(R)-methyloxirane, the comparison of ROA spectral signs by the experimental observation, the calculations based on Mulliken charges and the effective charges developed from bond polarizabilities. N shows inconsistency to the experimental observation.

Experimental wavenumber	Experimental ROA spectral sign	Calculated Δ_d sign by Mulliken charges	Calculated Δ_d sign by the effective charges as t varies from 0. 0133 to 0. 0160
1501. 0		−	+
1477. 0	−	−	−
1458. 0		+	−
1410. 0	+	+	+
1372. 0		+	+
1268. 0		−	+
1170. 0	+	N−	+
1147. 0		+	−
1134. 0	−	N+	−

1106. 0	+	+	+
1026. 0	—	N+	—
953. 0		—	+
898. 0	+	N—	+
832. 0	+	+	+
750. 0	—	N+	—

The final determination of the scaling factor t is of significance. It relates the relative bond polarizabilities to the electric charge. Furthermore, we note that, corresponding to $t=0.0133$ and 0.0160, there are 6.11 and 7.35($t \cdot \sum_l \partial\alpha/\partial R_l = t \cdot 459.64$) electrons being excited or involved in the vibronic coupling that leads to the Raman effect. In the methyoxirane molecule, there are 32 electrons, hence the excitation percentage or probability is between 19% and 23%. That is, around 20% of electrons are excited or perturbed in the Raman process under 532 nm excitation!

Comments

(1) The elucidation of the scaling parameter for equating bond polarizabilities to electric charges is significant. This is realized through fitting to the ROA signatures by our classical model. This is possible due to that ROA provides an extra condition in addition to what Raman offers.

(2) One may appreciate the finding that around 20% of electrons are excited or perturbed in the Raman process under 532 nm excitation by this simple/classical treatment of ROA, since this offers us a concrete impression on the Raman process.

9.5 The VCD application

For VCD of(+)-(R)-methyloxirane, the spectral signs of the VCD peaks by the experiment[4], the quantum chemical calculation by VCT-CO/6-31G* (0, 3)[4] and our method based on the Mulliken charges are listed in Table 9. 3. In fact, the disadvantage of our method is not serious as compared to the quantum chemical VCT-CO/6-31G* (0, 3) calculation, if one notes that by our algorithm based on the Mulliken charges, of the 12 modes, only the two modes at 1147 cm^{-1} and 898 cm^{-1} are of significant inconsistency, the other three discrepancies occur in the modes at 1501 cm^{-1}, 1170 cm^{-1} and 1106 cm^{-1} of rather small VCD intensities; while by the quantal method, there are two cases of discrepancy, one is at 1170 cm^{-1} of very small intensity and the other at 1134 cm^{-1} of larger intensity. The accuracy of the prediction of VCD spectral signs by our algorithm is noticeable. Indeed, for the VCD process, unlike that of ROA in which the charge excitation in the Raman virtual state is crucial, the Mulliken charges in the ground state simply play the major role.

Table 9. 3 The comparison of the VCD spectral signs of(+)-(R)-methyloxirane by the experimental[4], VCT-CO/6-31G* (0, 3)calculation[4] and our method based on the Mulliken charges. N shows inconsistency to the experimental observation.

Experimental wavenumber	Experimental VCD relative intensity and its sign	Sign by VCT-CO/6-31 G* (0, 3)	Sign by our method
1501. 0	11	+	N −
1477. 0			−
1458. 0			+
1410. 0	51	+	+
1372. 0	18	+	+
1268. 0	−30	−	−
1170. 0	6	N −	N −
1147. 0	−36	−	N +
1134. 0	53	N −	+
1106. 0	− 14	−	N +
1026. 0	15	+	+
953. 0	−100	−	−
898. 0	76	+	N −
832. 0			+
750. 0	35	+	+

Comments

Though roughly, it is conceivable that the core difference between ROA and VCD is that there is charge excitation/disturbance in ROA, but not in VCD. For a comprehensive study on Raman and ROA, this charge excitation/disturbance has to be taken into consideration seriously. This has been of our central concern and is the core point we have been emphasizing throughout this book!

References

[1] Barron L D. Molecular Light Scattering and Optical Activity. Cambridge: University Press, 1982.

[2] Freedman T B, Balukjian G A, Nafie L A. J. Am. Chem. Soc., 1985, 107: 6213.

[3] Freedman T B, Lee E, Nafie L A. J. Phys. Chem. A 2000, 104: 3944.

[4] Rauk A, Eggimann T, Wiester H, et al. Can. J. Chem., 1994, 72: 506.

Appendix A

One way to find out the bending coordinates that are more coupled to the stretching coordinates

In the bond polarizability analysis, we need to find out those vibrational modes that are more composed of stretching coordinates and those bending coordinates that are more coupled to the stretching coordinates. In this Appendix, we show one convenient way.

Suppose there are $S_1, \cdots, S_r$ stretching coordinates. We can arrange the vibrational modes in a descending order according their PED sums of these stretching coordinates. Then, we can choose the first r mode intensities for the bond polarizability analysis of these r stretching coordinates. Of course, these r modes may also contain the bending coordinates which are more strongly coupled to the stretching coordinates. The bending coordinates with larger PED in the first r modes and in those modes just after the first r modes are the potential ones that we may choose. By this way, we can determine, besides the first r mode intensities, how many more mode intensities are required in accordance with the number of the bending coordinates we might choose that are more strongly coupled to the stretching coordinates.

We take the case of (S)-(+)-2, 2-dimethyl-1, 3-dioxolane-4-methanol shown in Table 8. 3 as an example. We have the stretching coordinates, S1, ⋯, S9 (They are the C—C and C—O stretchings). The first 9 modes with the larger PED sums of these stretching coordinates in the descending order is:

Mode wavenumber	PED sum	PED
1058. 7	102	S9(100)
850. 7	86	S1(27) S2(35) S4(11) S6(10) S47(10)
1030. 1	72	S4(14) S5(32) S6(14)
1072. 8	71	S2(10) S3(58) S22(10)S40(12)
1095. 1	70	S4(13) S5(44) S23(14)
848. 5	70	S1(13) S4(34) S6(13)

793.7	58	S1(20) S2(21) S23(17)
968.8	53	S7(14) S8(10) S52(20)
1232.7	40	S8(27) S27(15) S46(21)

These 9 mode intensities can be employed for elucidating the bond polarizabilities of S1—S9. Meanwhile, we note from PED of these 9 modes that the 7 bending coordinates, S22, S23, S27, S40, S46, S52 and S47, are evidently strongly coupled to S1—S9. The modes after the first 9 modes arranged in the descending order by their PED sums of S1—S9 are:

Mode wavenumber	PED sum	PED
1277.6	35	S7(25) S27(11)S52(13)
924.8	33	S7(12) S8(20) S46(39)S52(23)
650.3	32	S7(19) S8(12) S22(34)
950.4	22	S35(36) S37(15)S40(25)
1170.4	16	S28(19)S47(24)S51(14)
750.1	12	S22(17) S23(37) S40(10)
507.5	10	S27(77)

Then, these 7 mode intensities, together with the previously chosen 9 mode intensities, can be employed for the elucidation of the bond polarizabilities of S1—S9 and S22, S23, S27, S40, S46, S52, S47.

In the phase determination, we usually do not set conditions on those of the bending coordinates. The increase of the adopted bending coordinates will result in more indeterminate phase sets. However, the variation among the results by these indeterminate phase sets are not serious, especially as far as the physical picture deduced therefrom is concerned. Moreover, we have already noticed that the variation in the differential bond polarizabilities by these indeterminate phase sets is particularly very small in many cases.

Appendix B
References for the work on Raman and ROA intensities

[1] Liu G, Wu G. Surface enhanced Raman study of benzidine on Ag electrode: an interpretation for different adsorption configurations at various applied voltages. J. Mol. Struct. , 1987, 161; 75.

[2] Tian B, Wu G, Liu G. The molecular polarizabilities and their implications as interpreted from the surface enhanced Raman intensities: a case study of piperidine. J. Chem. Phys. , 1987, 87: 7300.

[3] Tian B, Wu G. EHMO calculation of pyridine adsorbed on Ag adatom. International Journal of Quantum Chemistry, 1988, 33: 529.

[4] Liu G, Wu G. The Raman spectroscopic study of supersaturation: ν_1 mode of aqueous $NH_4H_2PO_4$ solution. Spectrochimica Acta, 1988, 44A: 1007.

[5] Huang Y, Wu G. Force constants and bond polarizabilities of thiocyanate ion adsorbed on the silver electrode as interpretated from the surface enhanced Raman scattering. Spectrochimica Acta, 1989, 45A: 123.

[6] Qi J, Wu G. Temperature dependent surface enhanced Raman spectroscopy of piperidine in AgBr sol. Spectrochimica Acta, 1989, 45A: 711.

[7] Huang Y, Wu G. Variation of bond polarizabilities of adsorbed pyrazine molecule on the silver electrode as a function of applied voltage from the surface enhanced Raman scattering. Spectrochimica Acta, 1990, 46A: 377.

[8] Wu G. Charge shift of adsorbed pyridazine molecules to the silver electrode surface at various applied voltages as interpreteted from the surface enhanced Raman intensities. J. Mol. Struct. , 1990, 238: 79.

[9] Zhong F, Sun X, Wu G. The bond polarizability derivatives of adsorbed methylviologen on the rough Ag electrode and their implications as interpreted from surface enhanced Raman intensities. J. Mol. Struct. , 1993, 298: 55.

[10] Zhong F, Wu G. The noncharge transfer mechanism of surface enhanced Raman scattering of pyrazine and methylviologen adsorbed on Ag and Au electrode via 1. 06 μm laser excitation: a study of bond polarizabilities as elucidated from the Raman intensities. J. Mol. Struct. , 1994, 324: 233.

[11] Ma S, Wu G. The surface enhanced Raman study of the charge transfer complex of pyridine-iodine on the silver electrode. J. Mol. Struct. , 1995, 372: 127.

[12] Ma S, Wu G. A classical theory for Raman optical activity. Chinese Phys. Letters, 1998, 15: 753.

[13] Wang P, Wu G. Ultraviolet laser excited surface enhanced Raman scattering of thiocyanate ion on the Au electrode. Chem. Phys. Letters, 2004, 385: 96.

[14] Liu Z, Wu G. Surface enhanced Raman scattering of thiourea adsorbed on the silver electrode: bond polarizability derivatives as elucidated from the Raman intensities. Chem. Phys. Letters, 2004, 389: 298.

[15] Wang H, Wu G. The monothiocyanate complexes of chromium ion(III) on the silver electrode by the surface enhanced Raman scattering. Spectrochimica Acta, 2005, A62: 415.

[16] Wang H, Wu G. Conformational rotation of 2-amino-1-butanol adsorbed on the Ag and Au electrodes as evidenced by the Surface enhanced Raman scattering. Chem. Phys. Letters, 2005, 407: 533.

[17] Liu Z, Wu G. Asymmetric environment around the thiourea molecule in aqueous solution as evidenced by the bond polarizability derivatives from Raman intensity. Chem. Phys., 2005, 316: 25.

[18] Liu Z, Wu G. The electro—oxidative activity of cysteine on the Au electrode as evidenced by surface enhanced Raman scattering. Spectrochimica Acta, 2006, A64: 251.

[19] Wang H, Wu G. The electronic structures of the nonresonant Raman excited virtual states of 2-aminopyridine by 632. 8 nm and 514. 5 nm excitations as evidenced by the bond polarizabilities. Chem. Phys. Letters, 2006, 421: 460.

[20] Zhao Y, Wang H and Wu G. The charge shift in the excited virtual state of pyrimidine during the nonresonant Raman process at 632. 8 nm: the bond polarizability study. Spectrochimica Acta, 2007, 66A: 1175.

[21] Fang C, Wu G. Electronic structures of nonresonant Raman excited virtual states: a case study of ethylene thiourea. J. Raman Spectroscopy, 2007, 38: 1416.

[22] Fang C, Liu Z and Wu G. A temporal study of the surface enhanced Raman bond polarizabilities of ethylene thiourea: the electromagnetic and charge transfer mechanisms. J. Mol. Struct., 2008, 885: 168.

[23] Fang C, Wu G. The significance of the temporal bond polarizabilty relaxation of 2- and 3-aminopyridine by 514. 5 nm excitation for the nonresonant Raman virtual states. Spectrochimica Acta, 2008, A71: 1588.

[24] Fang C, Wu G. The relaxations of temporal bond polarizabilities of methylviologen adsorbed on the Ag electrode by 514. 5 nm excitation: a Raman intensity study. J. Raman Spectroscopy, 2009, 40: 308.

[25] Fang C, Wu G. Temporal Raman polarizabilities of piperidine in liquid and on the Ag surface with electromagnetic enhancement. J. Mol. Struct., 2009, 938: 336.

[26] Fang C, Wu G. Raman intensity interpretation of pyridine liquid and its adsorp-

tion on the Ag electrode via bond polarizabilities. Spectrochimica Acta, 2010, A77: 948.

[27] Wang P, Fang Y and Wu G. Temporal electronic structure of the non-resonant Raman excited virtual state of p-nitroaniline by 514 nm excitation via bond polarizabilities. Chinese Physics B, 2010, 19: 113201.

[28] Wang P, Fang Y and Wu G. Raman excitation of (+)-(R)-methyloxirane and its origin of optical activity via bond polarizabilities. J. Raman Spectroscopy, 2011, 42: 186.

[29] Fang Y, Wu G and Wang P. The asymmetry of the differential bond polarizabilities in the Raman optically active (+)-(R)-methyloxirane and L-alanine. Chem. Phys., 2012, 393: 140.

[30] Fang Y, Wu G and Wang P. The charge excitation in the Raman process as correlated from a classical theory for Raman optical activity: the case study of (+)-(R)-methyloxirane. Spectrochimica Acta, 2012, A88: 216.

[31] Shen H, Wu G and Wang P. The chiral asymmetry of the anti-symmetric coordinates by the Raman differential bond polarizability study of S-phenylethylamine. Chinese Physics B, 2012, 21: 123301.

[32] Wang P, Wu G. The asymmetry of (−)α-pinene as revealed from its Raman optical activity spectrum. Chirality, 2013, 25: 600.

[33] Wu G, Wang P. A bond polarizability interpretation of the Raman optical activity intensity: the case study of (S)-1-amino-2-propanol. J. Spectroscopy and Dynamics, 2014, 4: 11.

[34] Wu G, Wang P. Stereo-structural implication by the differential bond polarizability: the ROA intensity study of chiral (*S*)-2-amino-1-propanol. Chirality, 2014, 26: 255.

[35] Shen H, Wu G, Wang P. Intra-molecular enantiomerism in (R)-Limonene as evidenced by the differential bond polarizabilities. Spectrochimica Acta, 2014, A128: 838.

[36] Wu G, Wang P. Intramolecular enantiomerism in (*S*)-(+)-2, 2-dimethyl-1, 3-dioxolane-4-methanol: The interpretation of Raman optical activity intensity. Chirality, 2015, 27:820.

[37] Shi J, Shen H, Zhang L, Wang P, Fang Y and Wu. G. Intramolecular enantiomerism as revealed from Raman optical activity spectrum. J. Rarnan Spectrosc., 2015, 42:1303.

These are for the work on the Raman intensities of phase transitions

[38] Lai X, Wu G. Temperature dependent Raman study of beta barium metaborate by Raman scattering. Spectrochimica Acta, 1987, 43A: 1423.

[39] Wu G, Yu J. Raman intensities of the lattice modes of $KH_{1-x}D_x F_2$ as the order parameter for phase transition. Spectrochimica Acta, 1988, 44A: 1457.

[40] Zhou Y, Wu G. The study of phase transition of KHF_2 doped with sodium ion with Raman intensities of the lattice modes as the order parameter. Spectrochimica Acta, 1993, 49A: 581.

[41] Ding X, Wu G. The critical Raman intensity behavior of the nitrate ν_1 mode and its thermodynamic interpretation for the phase transitions of crystalline NH_4NO_3. Spectrochimica Acta, 1995, 51A: 709.

[42] Ma S, Wu G. The fractal interpretation of the phase transition of the crystalline $NH_{4-x}D_xNO_3$ system as evidenced by the critical Raman intensity behavior of the nitrate ν_1 and ν_4 modes. J. Raman Spectroscopy, 1996, 27: 615.

[43] Wang H, Wu G. The behaviors of the vibrational modes of doped $(NH_4)_{1-x}K_xNO_3$ and $(NH_4)_{1-x}Na_xNO_3$ crystals: the Raman intensity study. J. Phys. Chem. Solids, 1998, 60: 129.

[44] Ma S, Wu G, Wang H. Raman intensity study of scaling properties of Na^+, K^+, Ag^+ and Pb^{2+} doped NH_4NO_3 crystals during phase transitions. J. Chem. Phys., 1998, 108: 7758.

[45] Li S, Ma S, Wu G. Scaling properties of the ν_1 Raman active nitrate mode in doped ammonium and potassium nitrate crystals. J. Chem. Phys., 1998, 109: 10311.